Geological Field Manual

- A practical guide for students and enthusiasts -

To Théodile of course

About the author: Wilfried Bauer (b. 1963) studied geology at the RWTH Aachen University and earned his PhD in 1995 with a study of the metamorphic basement of the northern Heimefrontfjella in Antarctica. He continued his research on the Precambrian of Dronning Maud Land (Antarctica) and the southern Urals (Russia) as a Principal Researcher at Aachen University.
From 2005 to 2007 he was a Senior Survey Geologist at the British Geological Survey and worked for its International Department on a World Bank funded project in Madagascar. From 2007 to 2014 he was head of the research and development section of an Australian exploration company in Madagascar. Since 2015 he is Associate Professor at the Department of Applied Geosciences of the German University of Technology in Oman.

Wilfried Bauer

Geological Field Manual

A practical guide for students and enthusiasts

With 38 Figures and 11 Tables

Bibliografische Information der Deutschen Nationalbibliothek: Die Deutsche Nationalbibliothek verzeichnet diese Publikation in der Deutschen Nationalbibliografie; detaillierte bibliografische Daten sind im Internet über www.dnb.de abrufbar.

Herstellung und Verlag: BoD – Books on Demand, Norderstedt
2015

Text and figures: CC BY-NC 4.0 (Full license text:
http://creativecommons.org/licenses/by-nc/4.0/legalcode)

ISBN 978-3-7386-0206-7

Content

Foreword 3
1. Rock-forming minerals 5
    3.1 Properties of minerals 5
    3.2 Important rock-forming minerals 10
        3.2.1 Silicates 10
        3.2.2 Non-silicates 18
2. Rock classification and description 23
    2.1 Rock fabric 24
        2.1.1 Structural features 26
        2.1.2 Structural features 29
    2.2 Igneous rocks 29
    2.3 Dykes 31
    2.4 Nomenclature of igneous rocks 32
    2.5 Sedimentary rocks 37
        2.5.1 Clastic sedimentary rocks 39
        2.5.2 Chemical sedimentary rocks 42
        2.5.3 Organic sedimentary rocks 44
        2.5.4 Limestone classifications 44
        2.5.5 Top/bottom criteria 46
    2.6 Metamorphic rocks 48
        2.6.1 Types of metamorphism 48
        2.6.2 Nomenclature of metamorphic rocks 53
3. Basic principles of structural geology 56
    3.1 Orientation of geological planes 56
    3.2 Faults and shear zones 57

3.3 Folds 62
3.4 Joints 67
3.5 Lineations 68
3.6 Unconformities 70
3.7 Analysis of geological measurements 71
4. Equipment 74
  4.1 Geological equipment 74
  4.2 Safety equipment 80
5. Geological Mapping 82
  5.1 Keeping a field log 83
  5.2 Topographic maps for geological field work 84
    5.2.1 UTM coordinate system 86
    5.2.3 Contour lines 88
    5.2.3 Declination und inclination 89
  5.3 Geological mapping criteria 91
    5.3.1 Outcrops 91
    5.3.2 Fieldstones 92
    5.3.3 Geomorphology 93
  5.4 Geological maps 93
  5.5 Measuring geological structures 94
  5.6 Symbols in geological maps 97
  5.7 Collecting samples 99
Appendix A: Schmidt Net 103
Appendix B: Chronostratigraphic Tables 104
Appendix C: Comparison Charts 107

Foreword

This manual was originally compiled as an in-house field guide of an exploration company with international staff from three continents. Geologists and trainees with very different backgrounds in applied, economic and structural geology needed a guideline for their fieldwork. Although similar, and more comprehensive, manuals already exist, they are not pocket-size and more suitable for the preparation of field work in the office. This small, light-weight booklet will fit in every pocket and can be carried even into difficult terrain. It is by no means a textbook but a mnemonic device; a formal geological education is required to understand the often very brief descriptions.

I would like to thank Rollo Desoutter for his review of the manuscript.

Further reading

- Rock descriptions:

FRY, N. 1997. *The Field Description of Metamorphic Rocks.* 110 pp., Wiley; Chichester.

THORPE, R.S. & BROWN, G.C. 1993. *The Field Description of Igneous Rocks.* 160 pp., Wiley; Chichester.

TUCKER, M.E. 2003. *Sedimentary Rocks in the Field.* 3rd Ed., 162 pp., Wiley; Chichester.

- Geological mapping and fieldwork:

BARNES, J.W. & LISLE, R.J. 2006. *Basic geological mapping.* 184 pp., Wiley; Chichester.

BERKMAN, D.A. 2001. *Field Geologists' Manual.* 4th Ed., 395 pp., Australasian Institute of Metallurgy, Monograph 9.

COE, A.L. (Ed.) 2012. *Geological Field Techniques.* 323 pp. Wiley-Blackwell.

FREEMAN, T. 2007. *Procedures in Field Geology.* 6th Ed., 95 pp. Blackwell; Oxford

McCLAY, K.R. 1987. *The Mapping of Geological Structures.* Geological Society of London Handbook Series, 161 pp., Open University Press; Milton Keynes.

PASSCHIER, C.W., MYERS, J.S. & KRÖNER, A. 1990. *Field Geology of High-Grade Gneiss Terrains.* 150 pp., Springer; Berlin, Heidelberg, New York.

1. Rock-forming minerals

A mineral is a naturally occurring solid with a characteristic chemical composition. Generally, the chemical composition is inorganic (a few exceptions are known) and a crystal structure, which is defined as the orderly spatial arrangement of atoms in the interior of a certain mineral, resulting in a crystal lattice. Different periodicities of the atomic or ionic arrangements in different directions of the lattice result in physical anisotropies. The chemical composition within a mineral species can be slightly variable which leads to mineral varieties such as gemstones. Spatially non-ordered mineral substances like opal are called amorphous or geloid.

Currently more than 3000 minerals are known (without varieties) but only 30 to 50 are so common that they can be called "rock-forming". The knowledge of this limited number of minerals allows identification of about 99 % of all rocks at the Earth's surface.

The percentage of a certain mineral species in the rock composition can be estimated. Minerals with a total amount with more than 20 % are major components, those of less than 20 % are called minor components. Minerals that make up less than 1 % of a rock's volume are called accessories.

3.1 Properties of minerals

- **Crystal habit**

The habit of a crystal species depends on the conditions during crystal growth. Surfaces which grow without disturbances by other minerals only depend on the geometry of the crystal lattice and develop characteristic faces, which is called the habit. This crystal lattice is defined by the regular distances between individual atoms or ions/ion groups and by the angles between them. The lattice defines the shape and angular relations of the outer surfaces of a crystal species. The angular relations between the faces remain constant but the development of individual faces is highly variable depending on the physicochemical conditions during the crystallization. This leads to different crystal shapes, even within the same mineral species. These shapes are called "prismatic", "acicular", "tabular", "dendritic" or "spheroidal". Irregular shapes are called "botryoidal", "massive", or "wiry" (native metals). The habit of a mineral species within a rock is generally uniform, however it may change between different rock types.

Minerals which can grow without disturbances and develop their characteristic habit are called euhedral or idiomorphic. Minerals which show outer surfaces which do not follow their characteristic habit due to disturbances by other minerals are called anhedral or xenomorphic.

- **Hardness**

Hardness is a measure of resistance against scratching or abrasion. A mineral species which scratches another mineral species is termed harder than the mineral species which can be scratched. For practical reasons, F. MOHS introduced 1812 a scale based on ten minerals. These minerals are sorted by increasing hardness in Tab. 1 together with some alternative test objects.

- **Cleavage**

Cleavage describes the way a mineral may split apart along various planes. In thin sections, cleavage is visible as thin parallel lines across a mineral. A good cleavage indicates that the bonding between the atoms in the lattice is in certain directions weaker than in others. The quality of cleavage is described with the terms:

perfect, good, imperfect, none.

Table 1: Mohs´ hardness scale

Hardness	Mineral	Alternative test materials
1	Talc	Pencil lead (graphite)
2	Gypsum	Finger nail
3	Calcite	Copper coin
4	Fluorite	
5	Apatite	Pocket knife (steel)
6	Orthoclase feldspar	Glass
7	Quartz	
8	Topaz	
9	Corundum	
10	Diamond	

Even if a mineral species has no cleavage and breaks along non-lattice-defined planes, the fracture plane can be indicative of a mineral. It is possible to distinguish several types of fractures like:

even, uneven, conchoidal, splintery, hackly.

- **Colour**

Minerals can have a distinctive own colour (idiochromatic), if a colour-giving element is part of the chemical composition (e.g. Mn gives rhodochrosite = $MnCO_3$ a pink colour). Another way to get coloured minerals is the inclusion of microscopic chromatic minerals or traces of chemical giving a mineral unusual colours (allochromatism). These colouring elements do not show up in the mineral's formula. Traces of Cr in corundum for example, give the gemstone variety ruby its red colour. Tiny rutile needles (TiO_2) in quartz result in a bluish colour. But also lattice defects due to radiation or missing atoms can lead to colour effects like the well known blue halite.

- **Streak**

Rubbing a mineral on an unglazed porcelain plate will produce a mineral powder which has a distinctive colour, often different from the mineral colour itself. A golden-yellow pyrite will have a greenish-black streak. The streak is diagnostic for many minerals with metallic or half-metallic lustre.

- **Lustre**

Amount and quality of light reflected by a mineral's surface can significantly vary. These variations are a results of different physical properties like absorption, colour interference, and optical refraction in the outermost layers of the lattice. There are only subjective terms to describe lustre like:

metallic, vitreous (like broken glass), *adamantine* (like a diamond), *greasy, silky, pearly,* or *resinous*. A mineral may have no lustre at all which is referred as dull or earthy.

- **Other features**

Several other features can be useful to identify a mineral:

- Specific gravity (given as g/cm^3): depends on the packing density of atoms in the lattice and the chemical composition of the mineral, i.e. the specific atom. Most rock forming minerals have a specific gravity between 2.5 and 4 g/cm^3.

- Transparency

- Odour (clay minerals) or taste (halite)

- Twinning: intergrowth of two or more crystals along regular lattice planes

- Radioactivity

- Fluorescence

- Magnetism

3.2 Important rock-forming minerals

3.2.1 Silicates

The most important group of rock-forming minerals are silicates. They form more then 90% of the rocks in the Earth's crust. The fundamental component is the very stable SiO_4-tetrahedron, in which four O^{2-} -ions with a tetrahedral shape surround a single Si^{4+} -ion in the central position. Neighboring SiO_4 tetrahedrons can share oxygen ions so that these tetrahedrons are combined as groups, chains, double chains, rings, sheets or frameworks.

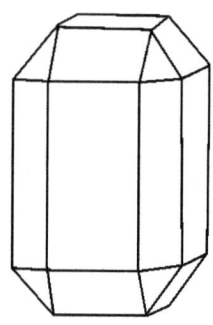

- **Olivine** $(Mg,Fe)_2[SiO_4]$

is a nesosilicate, which combines a single SiO_4^{4-} tetrahedron with variable amounts of the cations Mg^{2+} and Fe^{2+}. The Si:O-ratio is always 1:4. The pure Fe-end member is called Fayalite, the pure Mg-end member Forsterite.

Olivine forms translucent olive green, rhombohedral crystals without cleavage.

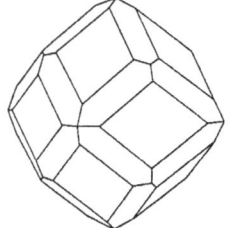

- **Garnet**

is a very common nesosilicate in metamorphic rocks. The general formula is $X_3Y_2[SiO_4]_3$ with X^{2+} = Mg, Fe, Ca, Mn and Y^{3+} = Al, Fe, Cr. The end members of these solid solutions have their own names. Colour and physical properties vary considerably with chemical composition. Most ubiquitous are reddish, multi-facetted garnet crystals which can look almost perfectly spheroidal in small individuals.

- Pyroxene

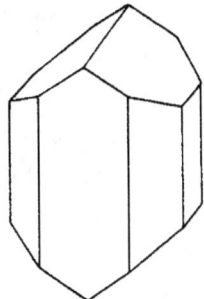

The name "Pyroxene" summarizes a group of structurally similar but chemically very variable minerals. The general formula is $XY(Si,Al)_2O_6$. They form chain silicates where a SiO_4 tetrahedron shares always two oxygen ions with its neighbors. The Si:O ratio is 1:3.

Pyroxenes are often green to greenish black, short prismatic crystals. They have two head faces and an octagonal prismatic outline. In sections perpendicular to the long axis an almost rectangular cleavage intersection pattern is a characteristic feature.

- Amphibole

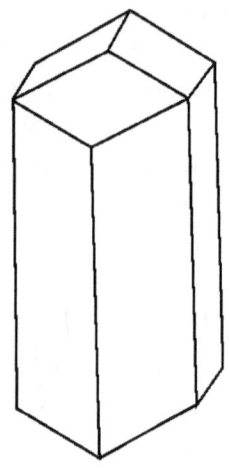

Also amphiboles are a group of structurally similar and chemically highly variable minerals. Always two SiO_4 tetrahedron chains are combined via connecting oxygen ions to a double chain, which results in a Si:O ratio of 1:2.75. Common amphiboles often form brownish green to black (Na-amphiboles also blue), longprismatic crystals. Euhedral crystals have three head faces and a hexagonal section. Sections show two distinctive sets of cleavage planes intersecting each other under an angle of 124°.

- Mica

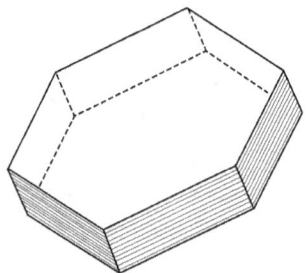

Micas are phyllosilicates where three of the four oxygen ions in a tetrahedron are shared with their neighbours, forming a plane sheet. The Si:O ratio is 1:2.5. Micas replace partially the tetravalent Si by trivalent Al and incorporate OH^- to balance the electric charge. The cations K^+ and Al^{3+} or K^+, Mg^{2+} and Fe^{2+}

connect the tetrahedral sheets. In the first case the resulting mineral is the colourless **muscovite**, in the second case the brown mica is called **biotite**. Micas have a perfect cleavage which allows them to be split into extremely thin, elastic sheets.

- **Clay minerals** and **chlorite**

Two other groups of phyllosilicates are characterized by alternating tetrahedron and octahedron layers. In clays the tetrahedral sheets are always bonded to octahedral sheets formed from small cations, such as aluminium or magnesium, coordinated by six oxygen atoms. The unshared vertex from the tetrahedral sheet also forms part of one side of the octahedral sheet but an additional oxygen atom is located above the gap in the tetrahedral sheet at the center of the six tetrahedra. This oxygen atom is bonded to a hydrogen atom forming an OH group in the clay structure. Clays can be categorised depending on the

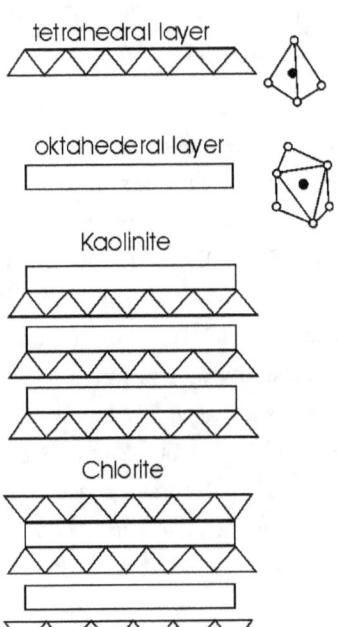

way that tetrahedral and octahedral sheets are packaged into layers. If there is only one tetrahedral and one octahedral group in each layer the clay is known as a 1:1 clay. The alternative, known as a 2:1 clay, has two tetrahedral sheets with the unshared vertex of each sheet pointing towards each other and forming each side of the octahedral sheet.

Chlorites have a 2:1 sandwich structure but unlike other 2:1 clay minerals, a chlorite's interlayer space is composed of $(Mg^{2+},Fe^{3+})(OH)_6$. Chlorite often appears as a pale green micaceous mineral but it lacks the elasticity of true micas. Clay minerals are generally less than 2 µm in diameter; they can only be identified by laboratory methods. Characteristic is the earthy odour.

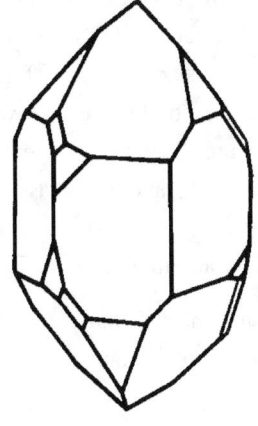

- Quartz SiO_2

Quartz has a framework lattice with a Si:O ratio of 1:2. All oxygen ions are connected to neighboring SiO_4 tetrahedra.

Quartz has no cleavage but the conchoidal fracture and the greasy lustre are good indicators. The mostly colourless quartz rarely shows a euhedral habit in rocks but well developed crystals are common in hydrothermal veins. Colour varieties are given discrete names. For example amethyst (violet), smoky quartz (brown), rose quartz (rose), citrine (yellow). Other varieties are amorphous (opal) or cryptocrystalline like the blood-red carnelian, white chalcedony or banded red to brown jasper.

- Feldspars

Feldspars are a group of tectosilicates with the three end members orthoclase ($KAlSi_3O_8$), albite ($NaAlSi_3O_8$) and anorthite ($CaAl_2Si_2O_8$). Solid solutions between albite and anorthite are

K-feldspar: Plagioclase:
Karlsbad twin polysynthetic
  twins

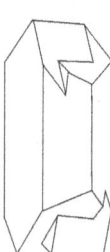 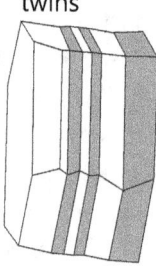

called **plagioclase**, solid solutions between orthoclase (**Potassium feldspar**) and albite are named alkali feldspar. The latter are stable under high-temperature conditions, at lower temperatures they exsolve to plagioclase and potassium feldspar. Anorthosite and orthoclase are immiscible.

All feldspars have good cleavages; in monoclinic K-feldspars after two rectangular oriented set of planes, in triclinic plagioclase almost rectangular set of planes. Twinning is very abundant, characteristic are Karlsbad twins in K-feldspar and polysynthetic twins after the albite law in plagioclases.

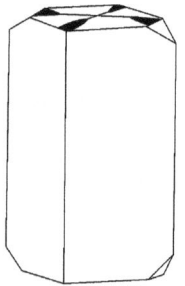

- **Andalusite, sillimanite, kyanite** $Al_2[O/SiO_4]$

This group of chemically identical but structurally different minerals is confined to metamorphic rocks. The orthorhombic andalusite occurs in low-pressure metamorphic rocks. Traces of carbon are often arranged in a cross pattern (see sketch), this variety is called chiastolite. All three minerals form long prismatic crystals. Indicative is the hardness anisotropy in the pale blue kyanite in different crystallographic directions. Sillimanite often occurs as translucent to white needles or fibres.

- **Serpentine** $Mg_6[(OH)_8/Si_4O_{10}]$

More than 20 different minerals form the serpentine group. They are commonly pale to dark green phyllosilicates with a characteristic silky lustre. Serpentine is formed by the alteration of olivine and therefore a common mineral in ultramafic rocks.

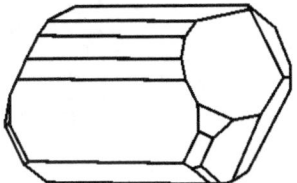

- **Epidote**

Epidote is the name for a group of Ca-Al-Fe^{3+} silicates crystallizing in the monoclinic system. Epidote is an abundant rock-forming mineral of secondary origin. It occurs in metamorphic or hydrothermally altered rocks. Crystals are short prismatic or isometric and have yellowish green to dark green colours.

3.2.2 Non-silicates

Amongst the non-silicates only carbonates (minerals with a CO_3 anion group) are an important component of the earth's crust. However, most natural concentrations of oxide, sulfide and halogenide minerals form important mineral deposits (ores).

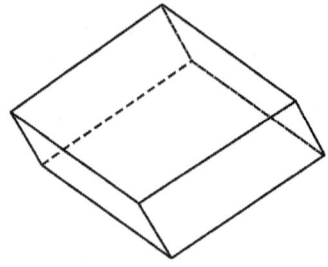

- **Calcite** $CaCO_3$

Is the most important rock-forming carbonate mineral. It has three equally perfect cleavages, which form perfect rhombohedra. With diluted hydrochloric acid (3%) calcite reacts after following equation:

$CaCO_3 + 2HCl \Leftrightarrow Ca^{2+} + 2Cl^- + CO_2 + H_2O$

The reaction is fierce and carbon dioxide escapes in bubbles from the acid-wetted surface.

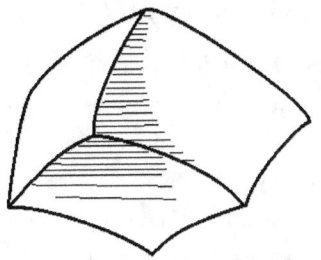

- **Dolomite** $MgCa(CO_3)_2$

Dolomite is formed from calcite by partial replacement of calcium by magnesium. Dolomite reacts slowly with diluted hydrochloric; the reaction can be recognized better with some mineral powder. Cleavage is much poorer than in calcite and typical are the curved crystal faces.

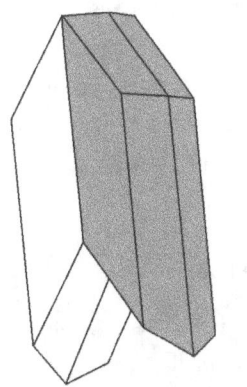

- **Gypsum** $CaSO_4 \cdot 2H_2O$

Gypsum is a calcium sulfate with two incorporated water molecules per formula unit. In large crystals three different cleavages, one of them almost as perfect as in micas, can be distinguished. Large slabs of transparent gypsum are named Maria-glass but more common are fibrous, silky aggregates. Twinning is very common, resulting in "swallow-tail" contact twins.

- **Hematite** Fe_2O_3, **Magnetite** Fe_3O_4 and **Limonite** $FeOOH$

These three minerals are the most common iron oxides and hydroxides. Hematite crystals are steel-grey with metallic lustre but their streak is cherry red. Thin hematite or hematite/limonite coats stain many terrestrial sedimentary rocks with a reddish or reddish-brown colour. Limonite occurs in yellow-brown earthy masses or as coats on other crystals. Magnetite is a common accessory in all rock types and can be easily identified due to its magnetic properties.

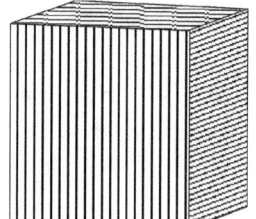

- **Pyrite** FeS_2

Pyrite typically occurs as euhedral small cubes of metallic yellow colour. The streak is greenish black. Very typical is the fine striation of the faces parallel to a crystal edge. Exposure to water and atmosphere transforms pyrite slowly to limonite and sulfate.

Diagnostic features of the most common rock forming minerals are summarized in Table 2.

Table 2: Diagnostic features of common rock-forming minerals.

Hard-ness	Common colours	Cleavage	Other features		Name		Habit
7	transparent, milky, smoky grey, others	none	resinous lustre, conchoidal fractures		Quartz		generally anhedral
6	red, white	2 planes good	vitreous, characteristic twinning		Orthoclase	Feld-spars	tabular, prismatic
	white, grey				Plagioclase		
3	silvery white	1 plane perfect	elastic flakes		Muscovite	Mica	flakes
	dark brown, black				Biotite		
2-3	green	1 plane perfect	non-elastic flakes		Chlorite		flakes
6	dark green, black	2 planes good	87° cleavage intersection	2 head faces	Pyroxene		long or short prismatic, amphibole can be acicular
			124° cleavage intersection	3 head faces	Amphibole		
7	dark red, brown, many other varieties	1 plane poor	isometric		Garnet		large crystals show a multifaceted euhedral habit, small crystals are spherical
6 ½	green	1 plane good, 1 plane poor	vitreous lustre		Olivine		granular
5–5 ½	steel grey, black, red	none	red streak		Hematite		granular, flaky, botryoidal

5-6	Yellow-brown, ochre	none	brown streak	**Limonite**	earthy masses, crusts
6-6½	metallic yellow	1 plane poor	greenish black streak	**Pyrite**	cubes
5½-6	metallic black	none	magnetic	**Magnetite**	cubes
3	white, grey, coloured	3 planes perfect	fierce reaction with diluted HCl	**Calcite**	granular, rhombohedral
1½-2	transparent, white, grey	2 planes perfect	swallow-tail twinning	**Gypsum**	granular, flaky, fibrous
1-2	white, grey	1 plane good	earthy odour	**Clay minerals**	flaky, mostly < 2μm

2. Rock classification and description

Rocks are classified by mineral and chemical composition, by the texture of the constituent particles and by the processes that formed them. These indicators separate rocks into igneous, sedimentary and metamorphic.

- **Igneous rocks**

All rocks crystallized at high temperatures from a molten magma either within the earth's crust or on the surface are called igneous rocks. Depending on the place of crystallization they are classified either as volcanic (surface) or plutonic (within the crust) rocks.

- **Sedimentary rocks**

Sedimentary rocks are formed by the deposition of rock fragments or minerals of eroded rocks and may sometimes contain significant amounts of organic debris. Components of dissolved minerals can under certain conditions be precipitated and form the subgroup of evaporitic sedimentary rocks with mineral textures equivalent to igneous rocks. Sedimentary rocks are deposited after their components have been transported by water, wind or glaciers over some distances.

- **Metamorphic rocks**

Metamorphic rocks are formed by subjecting any rock type (including previously-formed metamorphic rock) to different temperature and pressure conditions than those in which the original rock was formed. These temperatures and pressures are always higher than those at the Earth's surface and must be sufficiently high so as to change the original minerals into other mineral types or else into other forms of the same minerals by solid state recrystallization.

In a broader sense, migmatitic rocks which have undergone partial melting are also classified as metamorphic rocks.

Sedimentary and igneous rocks both occur as either hard or soft rocks. Sedimentary rocks are generally deposited as soft (incoherent) rocks in horizontal layers as widespread but relatively thin bodies. The transformation into a sedimentary hard rock is called diagenesis. Most igneous rocks are primarily hard rocks, forming bodies of irregular shapes; igneous soft rocks are less common and are related to explosive volcanism.

Soft rocks are primarily classified using grain size criteria (Tab. 5) whereas hard rocks are classified by their mineral composition and fabric.

2.1 Rock fabric

The geometrical and spatial configuration of all rock components is defined as the rock's fabric. A complete fabric description is part of every rock description and should be started in the field and completed with microscopical examinations later in the laboratory.

The term "texture" describes all geometrical properties of the components of a rock. This includes grain size, grain shape, habit and arrangement.

The term "structure" is used to describe the spatial relations of rock components.

Table 3: Rock description using textural features.

Crystallinity	Existing: holocrystalline				Non-existing: amorphous/ hyaline/glassy
Absolute grain size	very fine grained < 0,2 mm	fine grained 0,2-2 mm	medium grained 2-5 mm	coarse grained > 5 mm	
Relative grain size	inequigranular				equigranular
	porphyritic (plutonite)	porphyric (volcanite)	Porphyro- blastic (metamor- phite)	brecciated or conglomeratic (sedimentite)	
Shape of components	idiomorphic or euhedral hypidiomorphic xenomorphic or anhedral } igneous and metamorphic rocks			rounded angular special shapes } sedimentary rocks	
Cohesion of components	hard rock, directly or indirectly cohesive			soft rock, incohesive	

Table 4: Rock description using structural features.

Spatial arrange-ment					
	uniform	flow structure	slaty cleavage	metamor- phic foliation	bedding
Spatial filling	complete			incomplete: amygdaloidal, vesicular, porous	

2.1.1 Structural features

- Crystallinity: Rocks which have components crystallized from a melt or precipitated from a solution have a holocrystalline texture. Also metamorphic rocks, which are recrystallized under solid state conditions show a holocrystalline texture. Lack of macroscopic or microscopic visible minerals define a rock as hyaline (amorphous). Such natural glasses are volcanic rocks which cooled down before minerals had time to crystallize from the melt. Partially crystalline (crystals in a glassy matrix) rocks can be called semicrystalline. The term crystallinity should not be used for sedimentary rocks resulting from the deposition of inorganic or organic debris.

- Grain size: The first step is an estimation if main rock components are equigranular or not. If some components show significant variations in size, the rock is inequigranular which can be further described by distinctive names (Tab. 3). Plutonites with variation over a broad size spectrum are called porphyritic. Volcanic rocks with individual larger crystals in a fine matrix are called porphyric and rocks with coarser compents due to metamorphic growth are porphyroblastic.

Table 5: Classification of clastic sedimentary rocks by grain size.

Grain size of hard rocks	Soft rocks (DIN 4188)		Name (hard rock)	Old scientific names
coarse grained — 5 mm —	boulders	> 63 mm	Conglomerate Breccia	Psephite
	cobbles	63 - 20 mm		
	pebbles	20 – 6.3 mm		
medium grained — 2 mm —	granules	6.3 - 2 mm		— 2 mm —
fine grained — 0.02 mm —	coarse sand	2 – 0.63 mm	Sandstone Arkose Greywacke	Psammite
	medium sand	0.63 – 0.2 mm		
	fine sand	0.2 – 0.063 mm		
very fine grained	coarse silt	0.063 – 0.02 mm	Siltstone	— 0.02 mm —
	medium silt	0.02 – 0.006 mm		
	fine silt	0.006 – 0.002 mm		Pelite
	clay	< 0.002 mm	Mudstone	

The absolute grain size refers to the diameter of the components. Soft sedimentary rocks are classified by grain size. Several classification schemes exist, for technical purposes in Europe the German Standard DIN 4188 is widely accepted, in Anglo-American scientific literature the Udden-Wentworth scale is very common. For hard rocks a simple approach is used (Tab. 5).

- Shape of components: Minerals during early stages of crystallization from a melt or precipitation from a saturated solution often show their habit and are called idiomorphic or euhedral. During later stages, crystals are competing for the remaining space and are often disturbed in their growth by neighboring grains. Such crystals are refered to as anhedral or xenomorphic. Crystals showing partly idiomorphic and partly xenomorphic shapes are also called hypidiomorphic. In sedimentary rocks which are composed of debris from older rocks which are transported and mechanically abraded, transport distance and hardness determine their shape. They can be well rounded to angular. Sedimentary rocks mainly composed of organic components main still show the original shape (e.g. bivalves, corals, nummulites, etc.).

- Coherence: Individual components of a rock can be coherent (solid rocks) or non-coherent (soft rocks). If you have a hard rock, a further criterion is the direct or indirect contact of its main components. Indirect contact means that the individual grains are cemented by or embedded in a fine matrix. This is mostly the case for sedimentary rocks. The cement can be the same substance like the main components (quartz cement in a sandstone) or it can be another substance, precipitated from the pore water like calcite cement of a sandstone. In igneous rocks the "cement" can be a glass but generally most igneous rocks have direct contacts between the mineral grains. Their coherence is caused by the indentation and interfacial bonding.

2.1.2 Structural features

- Distribution and arrangement of components: Within a rock body, the distribution can be homogenous, which is common in large plutonic bodies. Due to regular material and/or grain size changes a rock body can show inhomogeneities. Reasons for such inhomegeneities in igneous rocks can be gravity segregation in the melt, orientation anisotropies due to melt flow or stress-related oriented crystallization (mostly in metamorphic rocks, => metamorphic foliation).

- Spatial filling: The spatial filling of a rock can be complete, i.e. no macroscopic visible pores or vesicles are visible, or is can be incomplete. Such incomplete filling results in porous, vesicular or amygdaloidal rocks. On a microscopic scale all rocks are more or less porous.

2.2 Igneous rocks

According to their geological origin, three subgroups of igneous rocks are subdivided: plutonites, volcanites and igneous dykes.

Plutonites are igneous rocks, which crystallize from melts within the earth's crust. These rocks solidify in depths between approximately 5 and 50 km. The generally low heat conductivity of rocks prevent melts from rapid cooling and so allow a slow, continuous crystal growth.

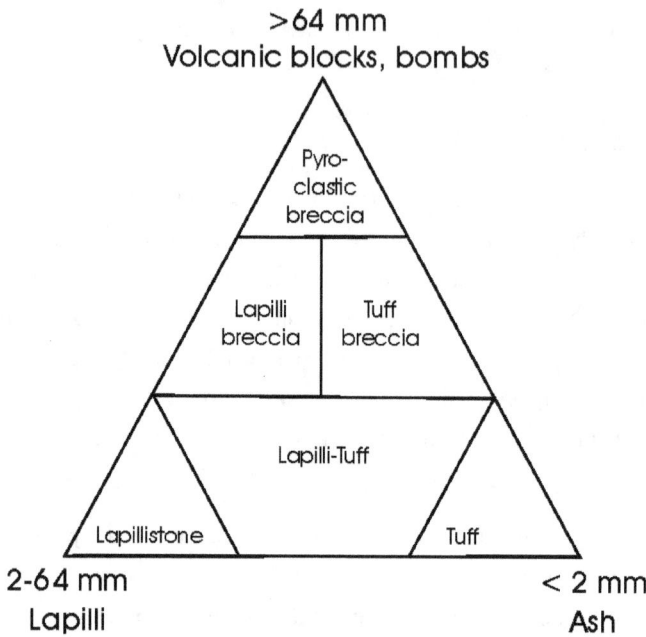

Figure 1: Classification of pyroclastic rocks.

Volcanites (effusive rocks) are formed by the rapid cooling of a silicate melt (lava) on the earth's surface under subaerial, submarine or subglacial conditions. Crystals formed under these conditions are either small (except being formed in the magma chamber earlier) or no crystals are formed and the rock is composed of silica glass. Crystals having been formed in the magma chamber can be significantly larger than the matrix.

Sudden pressure release of the gas dissolved in the melt will result in explosive eruptions. This violent eruption can also be triggered by contact with groundwater. The volcanic rock products so-formed are microvesicular, glassy and form small particles which solidify before being deposited. Fragmental material resulting from volcanic eruptions are called tephra (or pyroclastic rocks if fused together to hard rocks), regardless of composition and emplacement mechanism. The material is classified by grain size (Fig. 1) in ash, lapilli and volcanic blocks or bombs.

2.3 Dykes

Dykes are sheet-like intrusive bodies with high aspect ratios, which means that the thickness is much less than the other two dimensions. They intrude into fissures and frequently represent the pathways to feed a volcano (feeder dykes) or radiate from a feeder dyke into neighbouring host rocks. Spatially, steeply inclined dykes are distinguished from subhorizontal sills. Dykes originate from silicate magmas and can share the same composition with the volcanic or plutonic rock but the large contact planes to their host rocks lead to rapid cooling, resulting in smaller grain sizes. Such dykes are given the names of plutonic or volcanic rocks (see next section) prefixed by *micro*.

Another group of dykes show a different composition in comparison to their parental igneous rocks. They have distinctive chemical compositions and fabrics which do not occur in other igneous rocks. Dark dykes of this group are called lamprophyres (Tab. 6) and are classified by their plagioclase/alkali feldspar ratios, presence of foids (feldspathoids) and their mafic composition. Felsic dykes are generally classified by their grain size, since the chemical composition is generally dominated by quartz and K-feldspar but there is a wide variety of mafic components. Sugary, fine grained rocks are called aplites; coarse-grained rocks are pegmatites.

Table 6: Classification of lamprophyres

light components	alkali feldspar > plagioclase		plagioclase > alkali feldspar	
dark components	biotite, augite	hornblende, augite	biotite, augite	hornblende, augite
without foids	Minette	Vogesite	Kersantite	Spessartite
with foids		Sannaite		Camptonite

Sedimentary dykes are vertical bodies of sedimentary rock that cut off other rock layers. In cracks or large fissures infill of sediments form this type of dykes.

2.4 Nomenclature of igneous rocks

A simple, semiquantitative diagram, which allows the classification of igneous rocks is shown in Fig. 2. The rough average composition is shown as the length of a line beneath a certain rock name. There are three groups of names, the bottom row for plutonic names, the middle row for fresh volcanic rocks and the upper row for altered volcanic rocks. The top row gives the names for felsic and basic natural rock glasses. The SiO_2 content increases towards the left side, which results in a higher content of light minerals. Such rocks are called felsic, silicic or leucocratic. On the other side, rocks with dominating dark minerals, often SiO_2 undersaturated, are summarized with terms like mafic, basic or melanocratic. However, the small triangle on the left hand side labeled *Nepheline* also represents a silica undersaturated rock. Nephelin stands for a whole group of feldspathoids (sodalite, nosean, leucite and others), which never occur with quartz.

A formal quantitative classification was established by STRECKEISEN (1967) using Q-A-P-F double triangle diagrams (Fig. 3 a and b). **Q** stands for quartz, **A** for alkali feldspar (albite up to An_5 and K-feldspar, **P** for plagioclase (An_5 - An_{100}) and **F** for feldspathoids. Mineral percentages are normalized and then plotted in these diagrams. These diagrams are valid for all rocks with less than 90% mafic components. Rocks with more than 90% mafic minerals are plotted in special triangular diagrams (Fig.4 a-c).

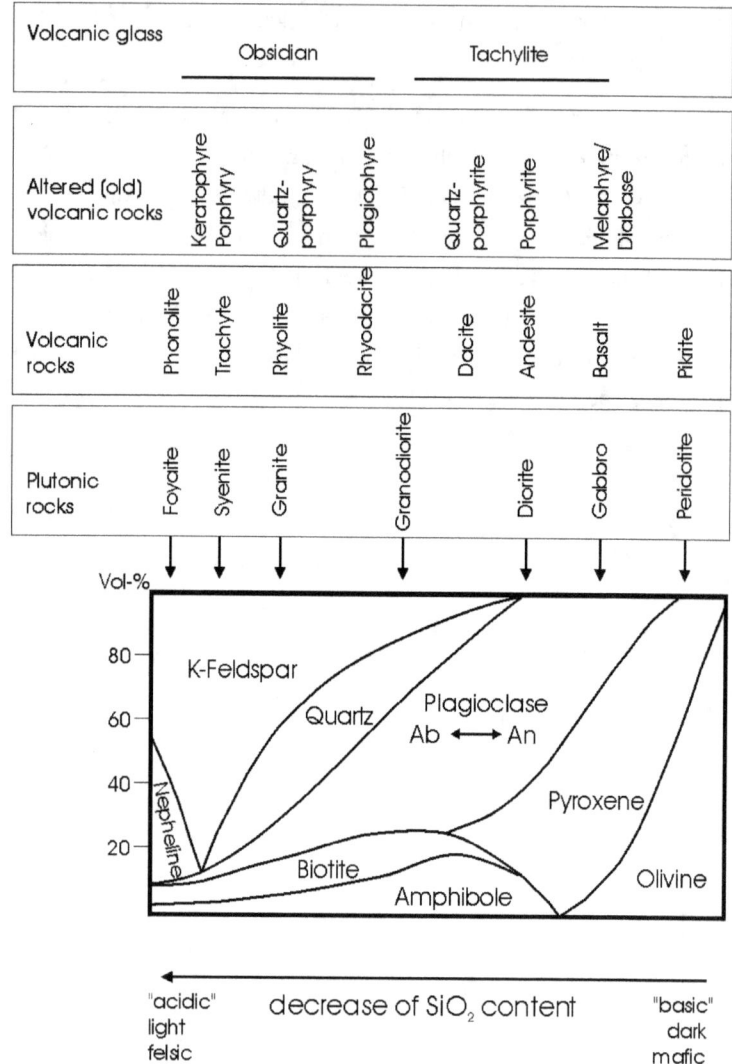

Figure 2: Semiquantitative diagram for the nomenclature of igneous rocks.

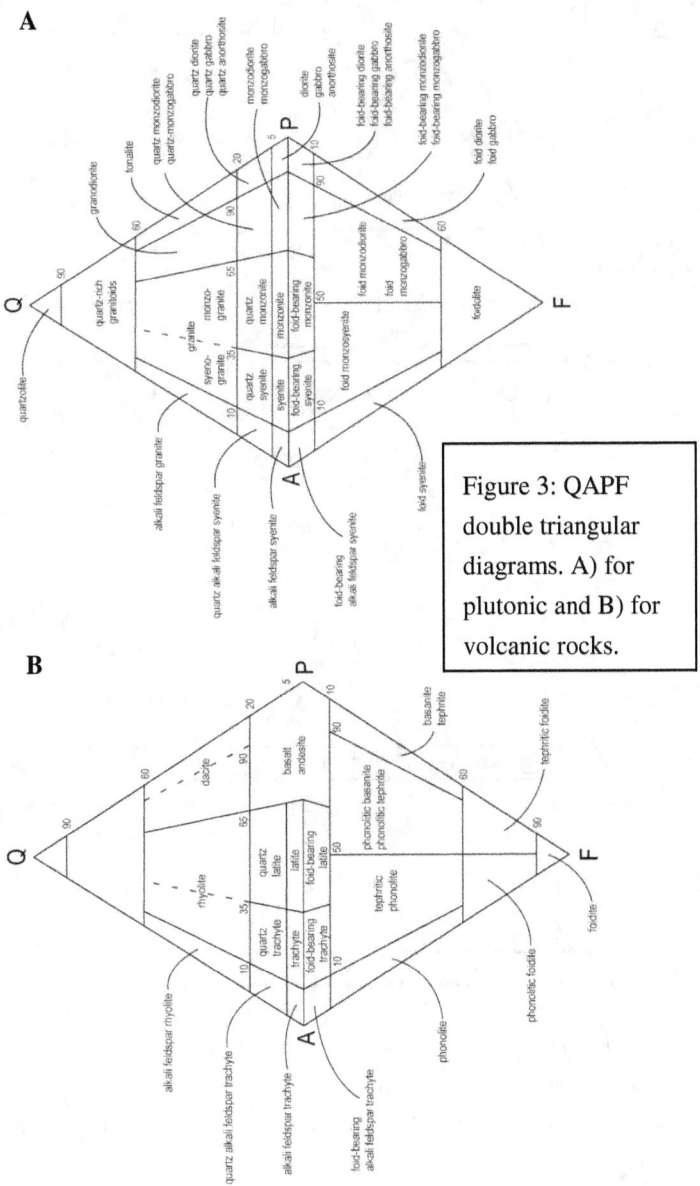

Figure 3: QAPF double triangular diagrams. A) for plutonic and B) for volcanic rocks.

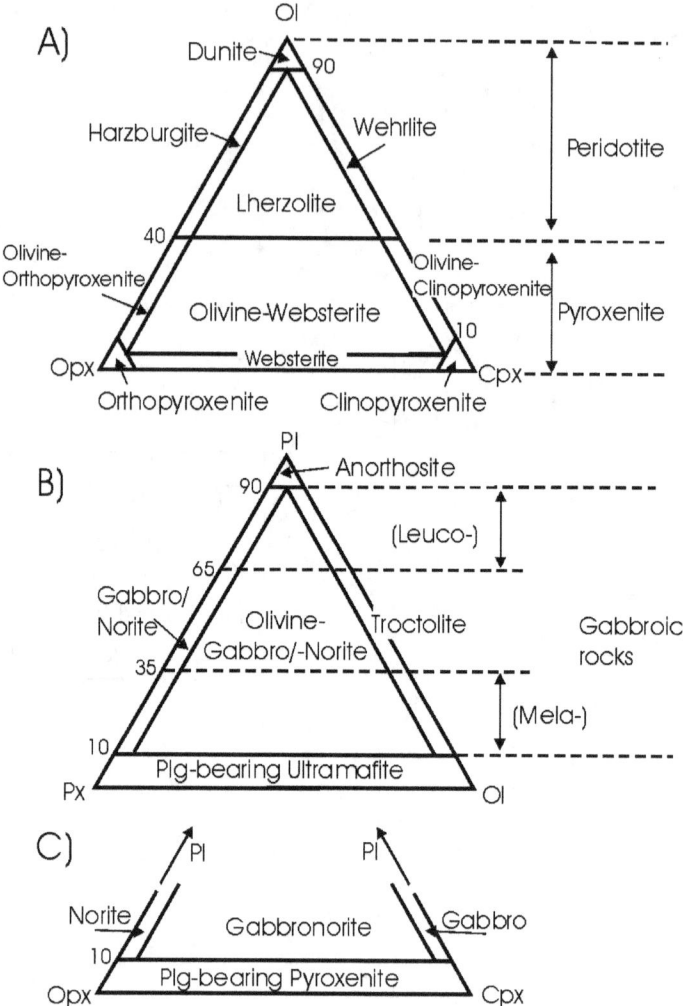

Figure 4 a-c: Triangular classification diagrams for ultramafic and gabbroid igneous rocks (MATTHES 1996). Abbreviations Ol=Olivine, Cpx=Clinopyroxene, Opx=Orthopyroxene, Px=Pyroxene in general, Plg=Plagioclase.

2.5 Sedimentary rocks

All rocks exposed on the Earth's surface undergo processes of chemical, physical and, to a minor degree, biological weathering. The weathering products are either rock fragments or dissolved material. Solid material can be transported by water, air, ice or mass movements and their deposition is related to a decrease of transport energy. Solutions require a liquid medium (generally water) and precipitate when the saturation level is exceeded. During transport and deposition organic components of plant or animal origin can be incorporated into sediments.

Table 7: Definition of different bed thicknesses.

cm		mm	
	very thickly bedded		very thickly laminated
100		3	
	thickly bedded		thickly laminated
30		1	
	medium bedded		medium laminated
10		0.3	
	thinly bedded		thinly laminated
3		0.1	
	very thinly bedded		very thinly laminated

Sedimentary rocks are deposited in strata that form a structure called bedding. Bedding is defined by the thickness of individual beds,

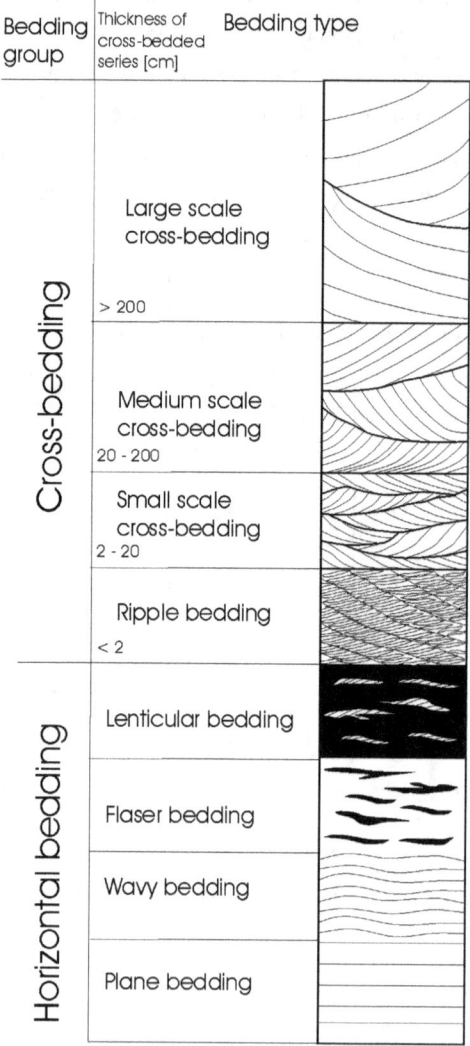

Figure 5: Common bedding types.

distinguishing between thick beds and thin laminae (Tab. 7) and internal features like cross-bedding (Fig. 5).

2.5.1 Clastic sedimentary rocks

Clastic sedimentary rocks are composed of fragments of pre-existing weathered and eroded rocks. Depending on transport medium and transport distance, individual fragments are rounded and sorted by their weathering and mechanical resistance. Immediately after their deposition, sediments are non-solidified soft rocks. They are classified by their grain size (Tab. 4). Later, such soft rocks can be solidified during a process called diagenesis.

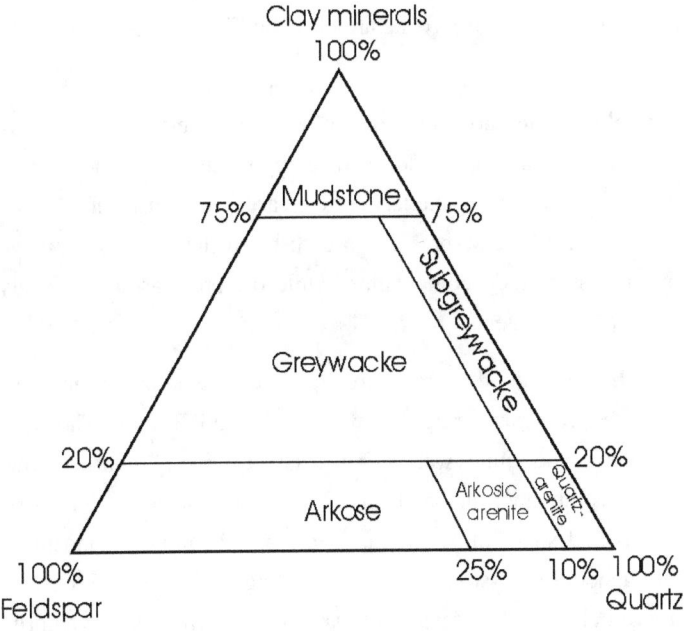

Figure 6: Triangular classification diagram for sandstones.

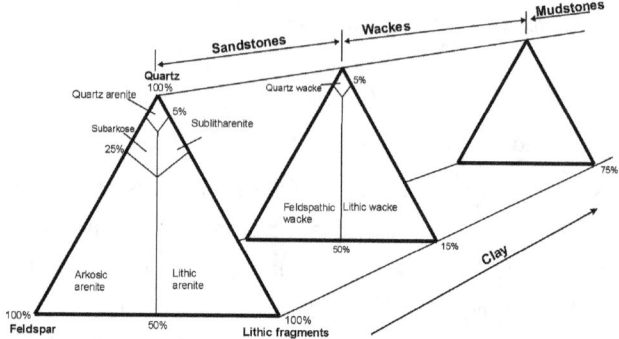

Figure 7: Classification of sandstones after PETTIJOHN et al. (1973).

Solid clastic sedimentary rocks are classified in four groups by their grain size:

- **Conglomerates and breccias** are coarse to very coarse, poorly sorted rocks mainly made up of components larger then 2 mm diameter. However, smaller particles are common in the matrix. Conglomerates and breccias are distinguished by the relative roundness of their components, while the breccias show mostly angular components.

- **Sandstones** (also known as arenites) are defined as clastic rocks with a grain size range between 0.06 and 2 mm. The very common and quite weathering-resistant mineral quartz is the main component in sandstones, but clay minerals, feldspars, micas and small lithic fragments are also abundant. Sandstones are generally classified in triangular diagrams where the corners represent quartz, feldspar and clay minerals (plus mica, chlorite and lithic fragments) as shown in Fig. 6 or in prismatic diagrams where the clay matrix is shown in the third dimension (Fig. 7).

- **Siltstone and shales** (also called pelites) are clastic sedimentary rocks with grain sizes not exceeding 0.06 mm. Pelites with grain sizes ranging from 0.06 to 0.002 mm are called shales or mudstones. Their mineralogical composition is very similar in comparison to sandstones. Loess, a weakly solidified aeolian sediment, was deposited during the last ice age on large areas in Europe and Central Asia. It contains up to 20% calcite. Rocks with grain sizes of less than 0.002 mm are predominantly made up of clay minerals, which can be formed by weathering of pre-existing silica minerals and/or during deposition and diagenesis.

- **Diamictites** are poorly or non-sorted rocks of different origin (Tab. 8), which as a general rule contain components of all sizes. Diamictites show no or only a very weak bedding.

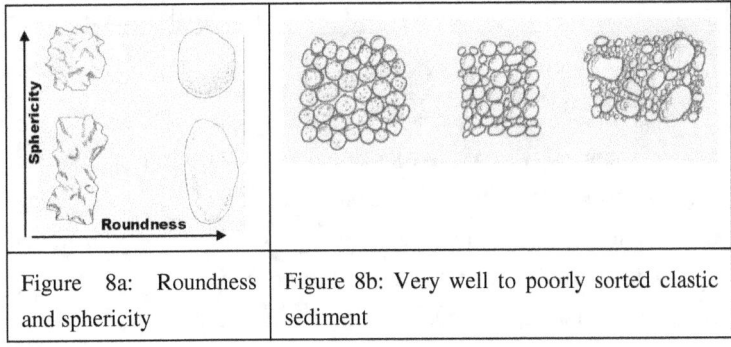

| Figure 8a: Roundness and sphericity | Figure 8b: Very well to poorly sorted clastic sediment |

Important criteria for the description of clastic sedimentary rocks are shape and sorting of their components. This allows some conclusions about transport distances and transport mechanism. The grain shape is defined as roundness and sphericity (Fig. 8a). With increasing transport distance, a grain tends to develop a spherical shape and the surface becomes rounder, i.e. smoother. Sorting depends on transport energy and viscosity of the transport medium. We distinguish between very well, well, medium, poorly, and very poorly sorted sediments or sedimentary rocks, considering the grain size distribution (Fig. 8b), but also the mineralogical composition. Well sorted sediments often contain a limited mineralogical spectrum with more than 98% quartz grains

Table 8: Criteria to distinguish different types of diamictites.

Type	Origin	Typical features
Fanglomerate	debris transported during episodically extreme rain events in semiarid areas	angular to semi-angular, often a red matrix
Tillite	consolidated moraine till	striated pebbles, often a dark matrix
Olistostrome	submarine debris flow	large boulders, chaotic internal structures

2.5.2 Chemical sedimentary rocks

During chemical weathering rock components become dissolved and are transported in solution by ground or surface water. In certain depositional environments like intracontinental basins or marginal basins temporarily separated from the open ocean, the water can become oversaturated due to high evaporation and minerals precipitate directly from the brine. The succession of evaporites precipitating from a marine brine is:

calcite – dolomite – gypsum – anhydrite – halite – potassium and magnesium salts

$CaCO_3$ (calcite) is the first product of the evaporite succession, however, precipitation is not the main process for the formation of limestones. More common than evaporation from a marine brine is the precipitation due to a shift in the bicarbonate equilibrium reaction ① from right to left:

① $\quad CaCO_3 + H_2CO_3 \leftrightarrow Ca^{2+} + 2(HCO_3)^-$

② $\quad CO_2 + H_2O \leftrightarrow H_2CO_3$

The direction of the equilibrium reaction ① depends on the carbon dioxide (H_2CO_3) concentration in the water, which produces bicarbonate, a weak acid. This reaction ② of atmospheric carbon dioxide (CO_2) with water (H_2O) proceeds to the left due to an increase of temperature, decrease of pressure or removal of carbon dioxide by biological processes. In these cases the acidity of the water decreases and reaction ① runs as a consequence to the left, leading to the precipitation of calcite.

In the tidal wave zone of lower latitudes calcite precipitates around crystallization nuclei (e.g. sand grains, small shell fragments) as concentric layers, due to the temperature increase of the water. The resulting spheres (ooids) remain in suspense until they are too large and heavy even in the zone of breakers. After settling down they can be solidified as a hard rock called oolite. A similar process leads to the formation of iron hydroxide oolite.

Other chemical sedimentary rocks result from a shift of the pH value (mostly in bottom or pore water). A decrease of the pH value for example due to decomposition of organic material, leads to an acidic environment which is favorable for the precipitation of silica gel ($SiO_2 \times nH_2O$), which can recrystallize to cryptocrystalline flint or chert.

2.5.3 Organic sedimentary rocks

Organic sedimentary rocks contain materials generated by living organisms, and include carbonate minerals created by organisms such as corals, bivalves and foraminifera, but also algae. Less common are silicic organic deposits created by sponges or unicellular organisms like diatoms. Other non-carbonatic deposits are phosphatic bonebeds. Deposits of higher land plants (peat, lignite, coal) and algae (oil shales) are the source of fossil fuels.

2.5.4 Limestone classifications

A simple classification by grain size distinguishes rudaceous (> 2 mm), arenaceous (2 mm - 0,063 mm) and lutaceous (< 0,063 mm) limestones in analogy to the grain size classification of clastic sediments. More common is a classification based on microtextural features according to FOLK (1959), however for a field description this classification is not suitable. DUNHAM (1962) introduced a classification which is based on the types of components larger than 2 mm and their ratio to the matrix. This classification (Tab. 9) can be used for field descriptions using a simple hand lens.

During diagenetic processes up to 50 % of the Ca^{2+} in calcite crystals can be substituted by Mg^{2+}, the resulting new mineral being called dolomite. A rock composed of dolomite crystals is also called dolomite or (less common) dolostone. Dolomites are pale yellow or yellow-brown rocks with a sugary texture. The replacement crystallization destroys the original limestone texture and most incorporated carbonate microfossils.

Table 9: DUNHAM Classification of limestones.

Autochthonous limestones (primary components bound by organisms during sedimentation/formation)	Boundstones	organisms constructed a calcareous framework Framestone			
		organisms encrusted components Bindstone			
		organisms acted as baffles Bafflestone			
Allochthonous limestones (primary components not bound during sedimentation)	more than 10% components > 2mm	grain supported Rudstone			
		mud supported Floatstone			
	less than 10% components > 2mm	without micrite	grain supported Grainstone		
		with micrite(< 0,03 mm)	Packstone		
			mud supported	> 10% grains Wackestone	
				< 10% grains Mudstone	

Marl or marlstone is the name for a rock containing a mixture of carbonate and non-carbonate minerals, the latter being mostly clay minerals. A classification of different mixture ratios is given in Tab. 10.

Table 10: Marlstone classification.

Limestone	95	80	65	% lime	35	20	5	Claystone
	Marly limestone	Limey marl		Marl	Clayey marl	Marly claystone		
	5	20	35	% clay	65	80	95	

2.5.5 Top/bottom criteria

After intense folding of sedimentary rocks the original stratigraphic sequence, i.e. top and bottom of individual beds can be reconstructed with a number of sedimentological or palaeontological criteria. Some examples of so-called geopetal fabrics are shown in Fig. 9.

In cross-bedded sedimentary rocks the geometry of the foresets are an important top/bottom criteria. The bottom part have a tangential contact with their base, whereas the top is truncated by the next bed (Fig. 9a). This criterion can be used for all types of cross-bedded

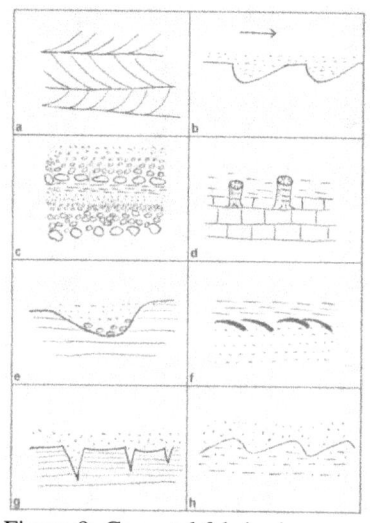

Figure 9: Geopetal fabrics in sedimentary rocks.

sediments, resulting from marine, fluviatile or aeolian deposition. Under very favorable conditions these features may be preserved even under low-grade metamorphic conditions.

Sole marks or flute casts (Fig. 9b) are sedimentary structures found on certain current-deposited strata like turbidites, which indicate small-scale current disruption, or scour. They are commonly preserved as depressions in the top of a rock bed (scour marks), or more commonly as casts of these indents on the bottom of the overlying bed (flute casts). Occurring as they do only at the bottom of beds, and having distinctive shapes, they make useful way up structures as well as palaeocurrent indicators.

In certain types of clastic sedimentary rock, the grain or clast size varies systematically from the base of the bed to its top. In a normally graded bed the grain or clast size is largest at the base and the bed is said to fine upwards (Fig. 9c). Beds deposited by density underflows such as turbidites typically show normal grading. Reverse grading or coarsening upwards is a characteristic of some alluvial fan deposits, so it is important to understand the depositional environment before using this particular criterion. Check the coarse beds for sole marks, they are common in normally graded (turbiditic) rocks.

Fossils in life position can be an important feature in Phanerozoic rocks. Typical examples are corals (Fig. 9d), sponges or bryozoans in marine environments and plants or their roots in continental environments. Also trace fossils and u-shaped burrows of non-fossilized suspension feeders can help to identify the way-up direction.

Channel fillings with a coarse-grained sediment at the bottom are typical features of fluviatile sediments (Fig. 9e). They cut into an underlying sediment and have only a limited lateral extension. Other features of continental sedimentary rocks that can be used as way-up indicators are mud cracks (Fig. 9g).

In shallow marine environments shells of bivalves are oriented with the convex side up (Fig. 9f). Asymmetric or symmetric ripplemarks occur in clastic or even carbonate sediments. Their shape in normal position typically shows a wide rounded base and a relatively sharp crest (Fig. 9h).

2.6 Metamorphic rocks

Metamorphism is defined as the change of the mineralogy of a pre-existing rock under solid state conditions in the Earth's crust. The temperature and pressure conditions are different from those prevailing during weathering and diagenesis and conditions which led to the formation of pre-existing rocks. Metamorphism is a broadly isochemical process with small scale chemical changes due to transport in fluids (e.g. H_2O, CO_2). Processes which involve significant changes in a rock's bulk chemistry are termed metasomatism.

2.6.1 Types of metamorphism

Metamorphic rocks are the result of the transformation of an existing rock type, the "protolith", at temperatures greater than 150 to 200 °C and pressures of 1500 bars causing profound physical change. The protolith may be sedimentary rock, igneous rock or another older metamorphic rock. Metamorphic rocks make up a large part of the Earth's crust and are classified by texture and mineral assemblage.

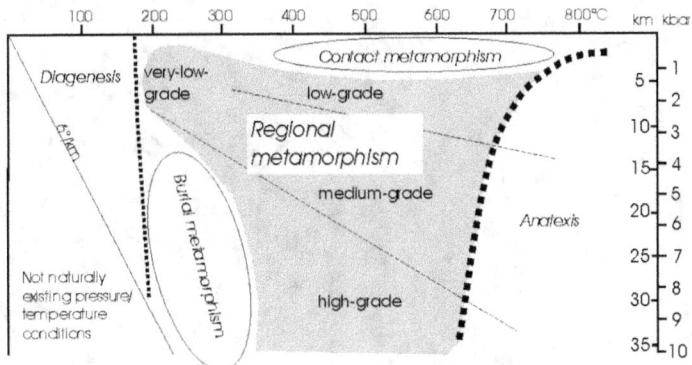

Figure 10: Schematic p-T diagram for different types of metamorphism (modified after WINKLER 1967).

The boundary between diagenesis and metamorphism is defined by the first appearance of some zeolite minerals (e.g. laumontite) which do not appear during diagenesis. This boundary is relatively pressure independent at a temperature of 180°C. The upper temperature limit of metamorphism is marked by the partial melting of rocks (anatexis). For rocks with a bulk granitic chemical composition and a water content of 5% anatexis starts at temperatures of 700°C (Fig. 10). In rocks with a different composition or a lower water content this boundary can shift significantly to higher temperatures.

Contact metamorphism occurs around hot igneous rocks as a result of thermally induced crystallization in the country rock. It also

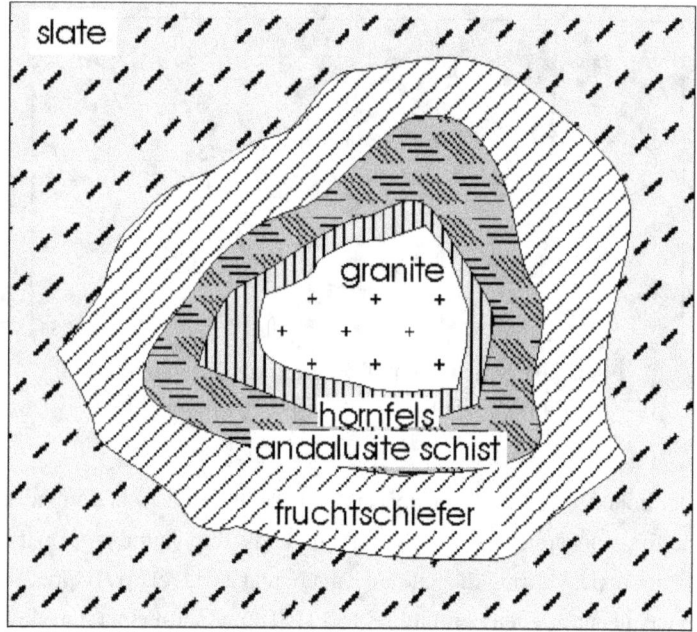

Figure 11: Schematic map of a granite pluton with its contact aureole.

affects xenoliths, incorporated into a magmatic body. The contact metamorphosed rim around an intrusive igneous rock is called an aureole (Fig. 11).

Burial metamorphism is a result of high lithostatic pressure at relatively low temperatures which is realized in the slab of a subducted oceanic plate. Such rocks can be exhumed and incorporated into an orogen where they mark ancient oceanic sutures.

Dislocation metamorphism is limited to narrow but extended zones of intense rock deformation due to stress. These zones are called shear zones.

At low temperatures near the crust's surface the resulting rocks are non-cohesive and not considered as "real" metamorphic rocks by many scholars, since they react only by fracturing of pre-existing minerals. All other rocks which show significant amounts of recrystallization are distinguished by their fabric and classified as cataclasites and mylonites (Tab. 11). Cataclasites are formed by brittle behavior with prevailing mechanical fracturing (= cataclasis), leading to a clast-rich, unfoliated rock. If these clasts are significantly larger than the matrix they are called porphyroclasts. Mylonites are formed in the field of ductile deformation and most minerals recrystallize to adapt their size to the local stress field. Recrystallization and resulting grain size depend on the strain rate and the temperature. At high temperatures and low strain rates minerals can even grow and such mylonites are coarser than their protoliths. At low temperatures and high stress rates olivine, pyroxene and amphibole in ultramafic rocks will recrystallize to chlorite, serpentine, talk and/or mica, and the resulting rock is a very fine-grained, schistose rock called phyllonite.

Table 11: Classification of shear zone-related rocks (modified after HEITZMANN 1985).

Non-cohesive rocks		Cohesive rocks (Dislocation metamorphic rocks s.str.)			
Kakirite	clasts	cataclasis > dynamic recrystallization	dynamic recrystallization > cataclasis	only dynamic recrystallization.	
		(grain size reduction)		(grain growth)	
Fault breccia	> 50%	Proto-cataclasite		Proto-mylonite	Blastomylonite
	10-50%	Meso-cataclasite		Meso-mylonite	
Fault gauge	< 10%	Ultra-cataclasite	Pseudo-tachylite	Ultra-mylonite	

Regional metamorphism covers large areas of continental crust typically associated with the deep seated roots of eroded orogens. Their products are affected by three factors, temperature, pressure and stress. This type of metamorphism represents mid- to lower-crustal regions. Under favorable conditions, anatexis leads to partial melting which is considered as the most intense form of metamorphism, resulting in migmatitic rocks.

Rocks become adapted to higher lithostatic pressures and temperatures by changing their mineralogical composition (mineral assemblages) and their fabric. The wide range of regional metamorphic rocks is subdivided into four zones (Fig. 10), named anchizone (very-low-grade), epizone (low-grade), metazone (medium-grade) and catazone (high-grade); typical mineral assemblages within these zones are also called metamorphic facies.

2.6.2 Nomenclature of metamorphic rocks

A generally accepted scheme for the classification of metamorphic rocks does not exist yet. For field descriptions, the following nomenclature concepts are permitted:

a) description based on the fabric and indicative minerals (phenotype),

b) description based on the protolith (genotype),

c) traditional names given by miners and stonemasons.

2.6.2.1 Phenotype

The phenotype describes the macroscopic visible structure of a metamorphic rock (Fig. 12):

- **Granofels**: massive metamorphic rocks without pervasive foliation, formed either under conditions without stress or a fabric typical for rocks composed of isometric minerals (without phyllosilicates). Typical for contact metamorphic rocks. Names can be specified by mineral qualifiers like spessartite-tremolite calcsilicate granofels.

- **Gneiss**: medium- to coarse-grained rock, widely spaced, often uneven metamorphic foliation; feldspar content > 20 Vol-%; Names are specified by mineral qualifiers: garnet gneiss, two-mica gneiss.

- **Schist**: fine-grained rock with closely spaced, smooth metamorphic foliation, feldspar content generally < 20 Vol-%; name specified by mineral qualifiers:

  Micaschist (muscovite and/or biotite as main components), greenschist (with chlorite and epidote)

  Slate: a rock at the transition between diagenesis and metamorphism; clay minerals are well oriented perpendicular to the main tectonic stress direction, mineralogical changes being within the lattice of the clay minerals, which can only be shown by X-ray diffraction methods.

- **Phyllite**: very fine-grained rocks, very high content of phyllosilicates, very smooth and extremely close-spaced foliation in the sub-millimeter range. Typical silky shimmer on the foliation planes.

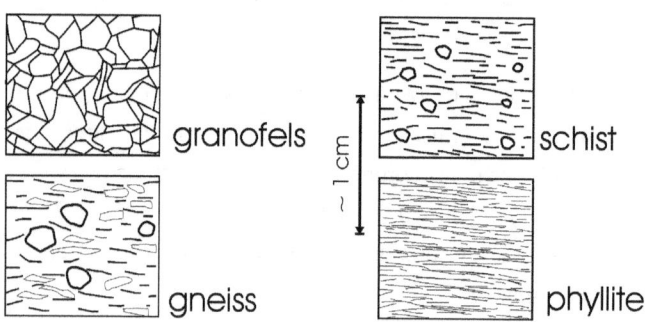

Figure 12: Classification of metamorphic rocks after their phenotype.

2.6.2.2 Genotype

If the protolith of a metamorphic rock can be identified, for example due to characteristic mineral assemblages or the setting in an exposure, a prefix is positioned in front of the name of the protolith or the phenotype:

Meta- metamorphic rock with known protolith, e.g. meta-granite, meta-greywacke

Ortho- metamorphic rock of igneous origin, e.g. orthogneiss

Para- metamorphic rock of sedimentary origin, e.g. paragneiss

2.6.2.3 Traditional names (selection)

Traditional names should be avoided since they have been used over time with different meanings and may not be unambiguous. On older geological maps a *quartzite* for example, may not be a metamorphic rock. Some modern trade names are also misleading (especially *marble*), so that such names should be avoided for scientific purposes whenever possible.

- Quartzite: Metamorphic rock with > 90% quartz (used instead of quartz granofels or quartz schist),
- Marble: Metamorphic rock with > 90% calcite (instead of Calcite granofels or Calcite schist),
- Serpentinite: Metamorphic rock with > 90% serpentine (instead of Serpentine schist),
- Amphibolite: Metamorphic rock with plagioclase and amphibole as main constituents,
- Eclogite: Metamorphic rock with garnet and omphacite (pyroxene variety) as main constituents,

- Granulite: Metamorphic rock with granitic composition but with garnet instead of biotite as main mafic phase.

For a metamorphic limestone following synonyms are permitted:

Meta-limestone = Calcite-granofels = Marble

3. Basic principles of structural geology
3.1 Orientation of geological planes

The orientation of a geological plane (sedimentary bed, metamorphic foliation, slaty cleavage etc.) in the three dimensional space is described by dip direction and dip angle (Fig. 13). Dip direction is measured 360 degrees counting clockwise from the North. The dip angle is measured from the horizontal downwards. The measured values are given as two numerals, the dip direction as a three digit number first, followed by a slash and the two digit dip angle.

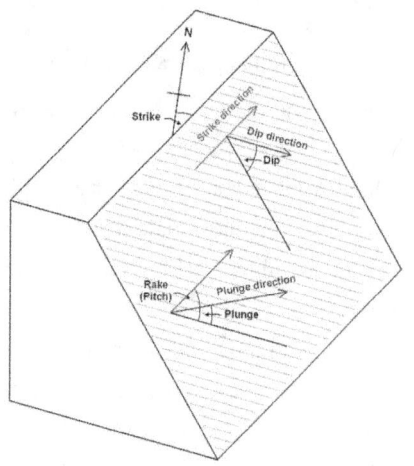

Fig. 13: Schematic presentation of dip direction, strike, dip angle of a plane and plunge of a lineation on that plane. (image from Wikipedia)

In older geological textbooks the measurement of strike and dip is used for the orientation of a plane but since strike has no polarity, the dip direction was added after the dip angle.

For example the orientation 120/45 means that the plane is dipping at an angle of 45° and in a direction of 120° clockwise from the North. In the old geological connotation with strike and dip, the same values were given as N30E/45SE.

3.2 Faults and shear zones

Rocks in the Earth's crust are affected by two principal types of deformation: compressional or tensional stress. Depending on the prevailing pressure and temperature conditions, the deformation rates and the strength of a rock type this stress will lead to strain and finally to mechanical failure.

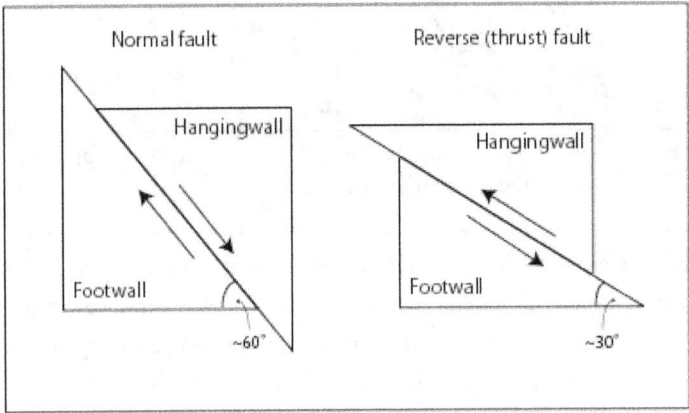

Figure 14: Schematic sections for the explanations of hangingwall and footwall. (image from Wikipedia)

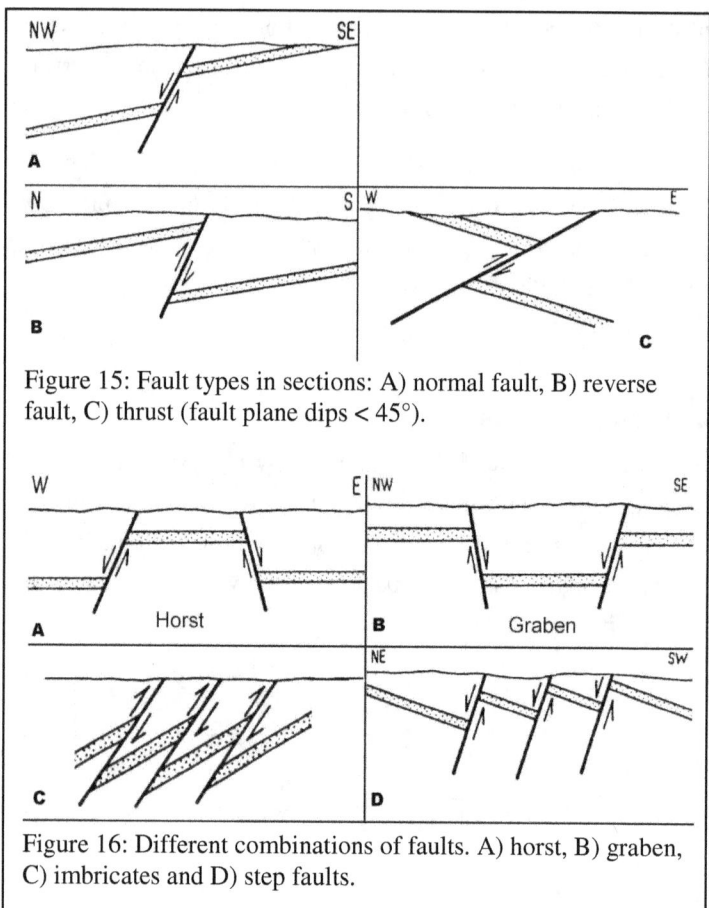

Figure 15: Fault types in sections: A) normal fault, B) reverse fault, C) thrust (fault plane dips < 45°).

Figure 16: Different combinations of faults. A) horst, B) graben, C) imbricates and D) step faults.

At temperatures below ~300°C mechanical failure occurs along planes which are called faults. At higher temperatures silicate minerals react by lattice gliding, twinning or grain boundary slip and the rocks show ductile deformation features. This rock type is formed in shear zones at deeper crustal levels. In contrast to a fracture plane, which generally ranges in thickness from 0 to a few meters, shear zones can be up to several kilometers thick. The basic principles of relative movements as shown in the next paragraphs are valid for both types of deformation.

The nomenclature of faults is defined by the relative motion of rocks on either side of the fault surface. Faults can be categorized into three groups based on the sense of movement or slip of the hangingwall (Fig. 14). A fault where the relative movement on the fault plane is approximately vertical is known as a dip-slip fault. Dip-slip faults can be again classified into the types reverse and normal. A normal fault occurs when the hangingwall moves downward with respect to the footwall (Fig. 15a). Alternatively such a fault can be called an extensional fault. A reverse fault is the opposite of a normal fault — the hanging wall moves up relative to the footwall (Fig. 15b). Reverse faults are indicative of shortening of the crust. The dip of a reverse fault is relatively steep, greater than 45°. A thrust fault has the same sense of motion as a reverse fault, but with the dip of the fault plane at less than 45° (Fig. 15c).

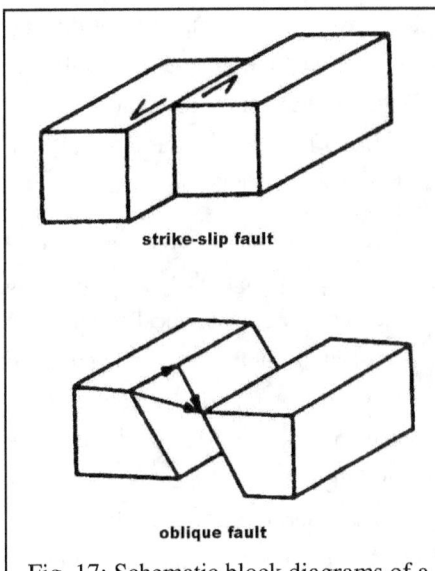

Fig. 17: Schematic block diagrams of a strike-slip fault and an oblique fault.

Some special names for the combination of fault planes exists. An upthrown block between two normal faults dipping away from each other is called a horst (Fig. 16a). A downthrown block between two normal faults dipping towards each other is called a graben (Fig. 16b).

The combination of several parallel normal faults is named step faults (Fig. 16d), whereas the combination of parallel reverse faults are called imbricates or, if they have low dip angles, imbricate thrust fans (Fig. 16c).

The fault surface is usually near vertical and the footwall moves either left or right or laterally with very little vertical motion. Strike-slip faults with left-lateral motion are also known as sinistral faults (Fig. 17 upper sketch). Those with right-lateral motion are also known as dextral faults. A fault which has a component of dip-slip and a component of strike-slip is termed an oblique-slip fault (Fig. 17 lower sketch). Nearly all faults will have some component of both dip-slip and strike-slip, so defining a fault as oblique requires both dip and strike components to be measurable and significant.

During field work, geologists have to pay special attention to faults and shear zones since the normal stratigraphic succession is disturbed by an offset. In particular, if an economically important reef, mineral vein or seam is offset by a fault, it is essential to determine the sense of movement on that fault and reconstruct the offset width in order to locate the continuation of the disturbed rock body. Such a fault has to be carefully mapped either on the Earth's surface or in galleries and adits. The determination of a brittle fault is possible with slickensides. The term refers to a smoothly polished surface caused by frictional movement between rocks along the two sides of a fault. This surface is normally striated in the direction of movement. The plane may be coated by mineral fibres that grew during the fault movement, known as slickenfibres, which also show the direction of displacement. Due to irregularities in the fault plane exposed slickenfibres typically have a stepped appearance that can be used to determine the sense of movement across the fault. The surface feels smoother when the hand is moved in the same direction that the eroded side of the fault moved.

Determining the displacements that occur in ductile shear zones is dependent on correctly determining the orientations of the finite strain axis. It is common practice to assume that the deformation is a plane strain simple shear deformation. This type of strain field assumes that deformation occurs in a tabular zone where displacement is parallel to the shear

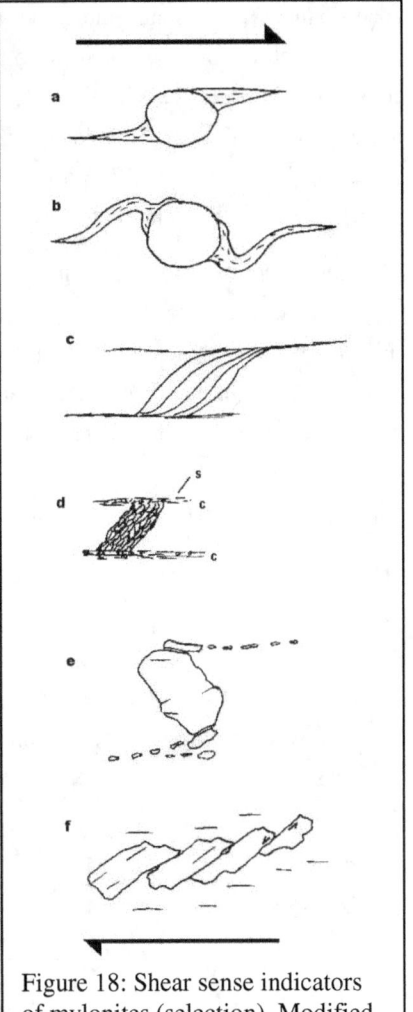

Figure 18: Shear sense indicators of mylonites (selection). Modified after Eisbacher (1996).

zone boundary. Kinematic indicators are structures in ductile shear zones that allow the sense of shear to be determined. Because of the constraints imposed by simple shear, displacement is assumed to occur in the foliation plane in a direction parallel to the mineral stretching lineation. Therefore a plane parallel to the lineation and perpendicular to the foliation is viewed to determine the shear sense.

The most common shear sense indicators are asymmetric σ porphyroclasts (Fig. 18a), δ porphyroclasts (Fig. 18b), and mica fish (Fig. 18c). These indicators are common in fine-grained, micaceous rocks. C/S fabrics (Fig. 18d) and mantled or fragmented porphyroclasts (Fig. 18e, f) are more common in coarser gneisses. All of these indicators have a monoclinic symmetry which is directly related to the orientations of the finite strain axis.

3.3 Folds

The term fold is used when a stack of originally planar surfaces are bent as a result of plastic deformation without mechanical failure of the rocks.

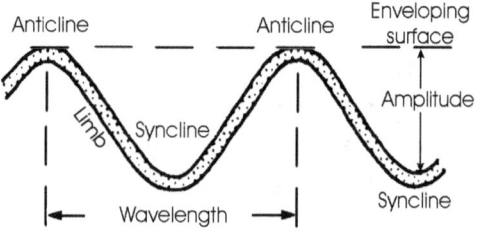

Figure 19: Geometrical elements of a fold.

Folds are described by geometrical elements. In Fig. 19 a simplified sketch shows a single folded bed. The folds have two different hinge portions and limbs. The hinge points are the points of minimum curvature radius of a fold.

The crest is called the anticline, the trough is called the syncline. On the limbs the inflection point marks the change of concavity. The distance from crest to crest or trough to trough is the fold's wavelength, the vertical distance from crest to trough the fold's amplitude.

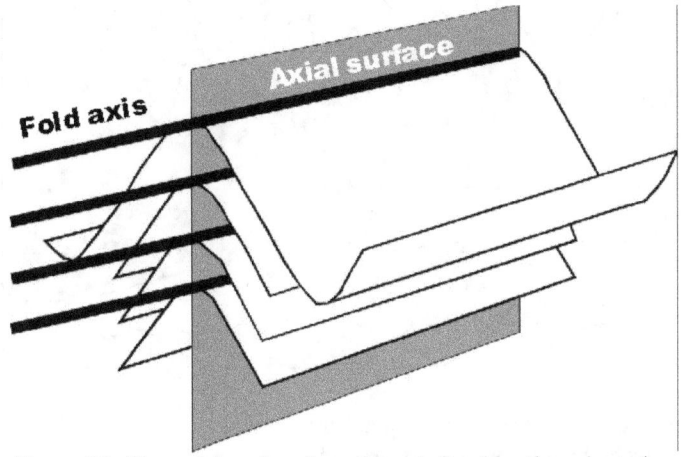

Figure 20: The axial surface is a plane defined by the orientation of the fold axes of individual beds.

Looking at a fold in the third dimension (Fig. 20), the hinge points along an entire folded surface form a hinge line, which is the fold axis. The trend and plunge of a linear fold axis gives you information about the orientation of the fold (Fig. 22). To completely describe the orientation of a fold, one must use the axial surface. The axial surface is the surface defined by connecting all the hinge lines of stacked folding surfaces. The axial trace is the line of intersection of the axial surface with any other surface (ground, side of mountain, geological cross-section).

The position of the axial surface is parallel to the cleavage and indicates the direction of maximum principle stress during the deformation. If the stress is directed horizontal with respect to the Earth's surface, the result will be vertical axial surface (and cleavage) and the folds are upright. A change of the principle stress

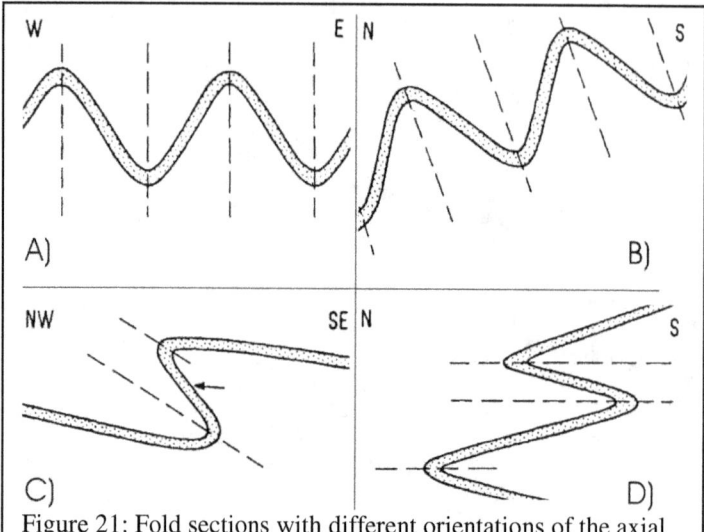

Figure 21: Fold sections with different orientations of the axial surface. A) upright folds; B) north verging folds; C) strongly vergent fold with one overturned limb; D) recumbent fold.

direction will result in an inclined axial surface and the axial surface will become inclined (Fig. 21). The inclination or vergence can increase with time until the axial surface is horizontal. The resulting fold is recumbent and way-up criteria are necessary to reconstruct the original stratigraphic sequence.

The fold description includes following criteria:

- Size: Microfolds are the smallest folds from microscopic to hand specimen size. Small scale folds range from hand specimen to outcrop scale, but they are still beyond a size which can be shown on large scale geological maps. Large folds are all folds which can be shown on geological maps.

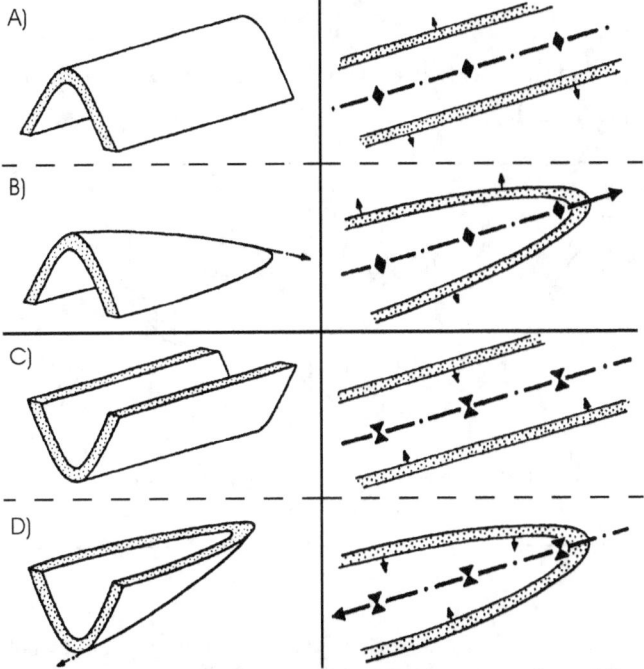

Figure 22: Examples of folds with different orientation of the fold axis; left side schematically shown with a single layer in three dimensions, right as the equivalent shape on a map. A) Anticline with horizontal fold axis; B) Anticline with a plunging axis; C) Syncline with a horizontal fold axis; D) Syncline with a plunging fold axis.

- Symmetry: Not all folds are equal on both sides of the axis of the fold. Those with limbs of relatively equal length are termed symmetrical, and those with highly unequal limbs are asymmetrical (Fig. 23). In some textbooks asymmetrical folds are also called S or Z folds. Looking downward the fold axis, the limbs will show the shape of the letter S or Z. In Fig. 21c an example of an S fold is shown, assuming we are looking down the fold axis.

- Position of the fold axis: the axis can be either horizontal or plunging (Fig. 22).

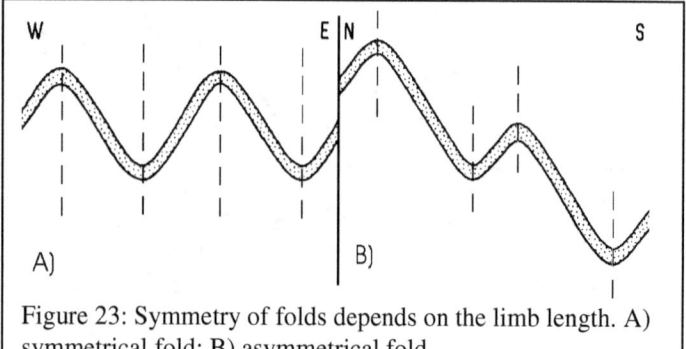

Figure 23: Symmetry of folds depends on the limb length. A) symmetrical fold; B) asymmetrical fold.

- Position of the axial surface can be upright, vergent (=inclined) or lying (Fig. 21); the vergence is indicated as the opposite direction of the axial surface's dip direction.

- Fold shape: some typical shapes are shown in Fig. 24. Most common are circular, cuspate and chevron folds in sedimentary rocks. Isoclinal folds require a high strain and are more common in metamorphic rocks.

- Fold tightness describes the opening angle between the limbs. Interlimb angles between 180° and 120° are called gentle, 120° to

70° open, 10°-30° close, <30° tight. Folds with parallel limbs are isoclinal.

- Deformation style classes: Folds which maintain uniform layer thickness are classed as concentric folds; those which do not are called similar folds. Similar folds tend to display thinning of the limbs and thickening of the hinge zone. Concentric folds are caused by warping which results from active buckling of the layers, whereas similar folds usually form by some form of shear flow where the layers are not mechanically active.

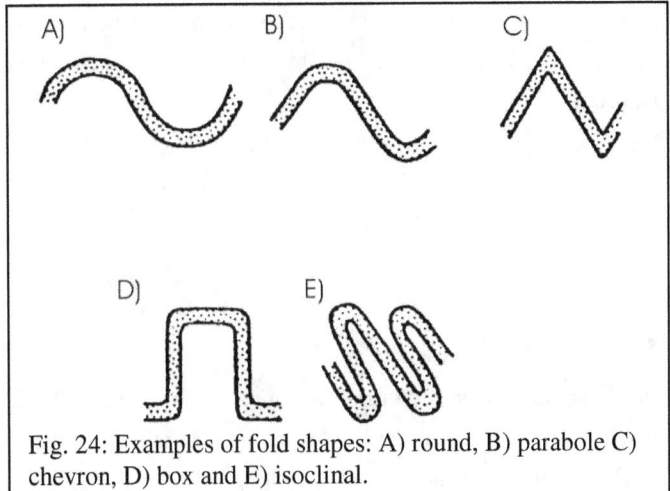

Fig. 24: Examples of fold shapes: A) round, B) parabole C) chevron, D) box and E) isoclinal.

3.4 Joints

The term joint refers to a fracture in rock where there has been no lateral movement in the plane of the fracture of one side relative to the other. Joints normally have a regular spacing related to either the mechanical properties of the individual rock or the thickness of the layer involved. Joints generally occur as sets, with each set consisting of joints sub-parallel to each other. Joints form in solid, hard rock that is stretched such that its tensile strength is exceeded. When this happens the rock fractures in a plane parallel to the maximum principal stress and perpendicular to the minimum principal stress. This leads to the development of a single sub-parallel joint set. Continued deformation may lead to development of one or more additional joint sets. The presence of the first set strongly affects the stress orientation in the rock layer, often causing subsequent sets to form at a high angle to the first set.

Joints form one of the most important types of discontinuity within rock masses, typically having no tensile strength. Therefore their quantification is an important task for engineering geologists. Length, frequency, orientation and opening width are parameters which have to be recorded on three walls, in ideal cases each perpendicular to the other.

3.5 Lineations

Lineations are linear structural features within rocks. There are several types of lineations, intersection lineations, crenulation lineations, mineral lineations and stretching lineations being the most common. Lineations are measured as lines with a plunge and a plunge azimuth.

Intersection lineations are linear structures formed by the intersection of any two surfaces in a three dimensional space. The trace of bedding on an intersecting foliation plane commonly appears as colour stripes generally parallel to local fold hinges. Intersection lineations can also be due to the intersection of two foliations.

Intersection lineations are measured in relation to the two structures which intersect to form them. For instance, original bedding, intersected by a fold axial plane foliation or cleavage, forms a δ intersection lineation, with an azimuth and plunge defined by the fold. This is the typical cleavage-bedding intersection angle and is diagnostic of the plunge of the fold on all parts of the fold. If two sets of cleavage intersect, the lineation is called ε intersection lineation.

Figure 25: Two sets of almost perpendicular crenulation cleavages in an anchimetamorphic schist. The vertical set overprints the horizontal set and must therefore be younger.

Stretching lineations are formed by shearing of rocks during asymmetric deformation of a rock mass. Stretching lineations record primarily the vector of greatest stretch, which is perpendicular to the principle plane of shortening. During this process elongated minerals such as amphiboles for example, are re-oriented in the direction of greatest stretch. Another process is the elongation of originally isometric mineral grains into rod-shaped individuals due to intracrystalline gliding. This is often the case in quartzite mylonites.

Crenulation lineations indicate an early stage of a cleavage formation. They occur either on the bedding planes of slaty rocks or on a cleavage plane as an embryonal second cleavage (Fig. 25).

3.6 Unconformities

An unconformity is an ancient erosion surface separating two rock masses of different ages, indicating that sediment deposition was not continuous. In general, the older rocks have been exposed to erosion for an interval of time before deposition of the younger rocks took place. The interval of geologic time not represented between the deposition of the older and younger rock units is called a hiatus. There are three principal types of unconformities: disconformities, nonconformities and angular unconformities.

Disconformity is an unconformity between parallel layers of sedimentary rocks which represents a period of non-deposition. An erosional disconformity is a special type where local erosion, for example by rivers, has carved out channels that are filled by younger sediments.

A **nonconformity** separates metamorphic or igneous rocks from overlying sedimentary rocks. These sedimentary rocks were deposited on an erosional surface that developed on metamorphic or igneous rocks.

Angular unconformity is an unconformity where horizontally parallel strata of sedimentary rock are deposited on tilted and eroded layers, producing an angular discordance with the overlying horizontal layers (Fig. 26). The whole sequence may later be deformed and tilted by further orogenic activity.

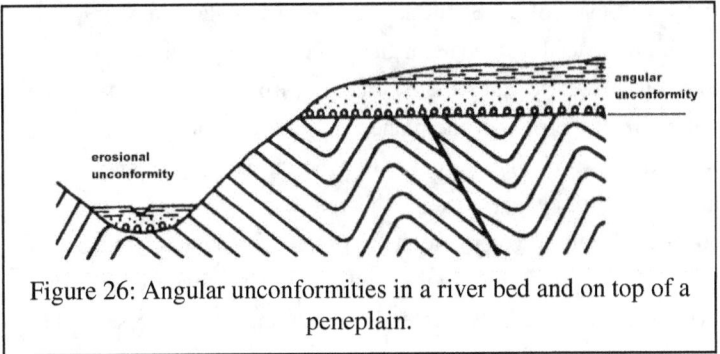

Figure 26: Angular unconformities in a river bed and on top of a peneplain.

3.7 Analysis of geological measurements

Large numbers of tectonic measurements can be analyzed using a stereographic projection like the Schmidt net. A Schmidt net is the Lambert azimuthal projection of a lower hemisphere. An example for copying is given as Appendix A. The top of the margin represents N, then counting counter clockwise we get all directions between 0 and 360°. The interior of the net represents the third dimension. An individual plane is either represented by a great circle or the pole on the great circle (Fig. 27). A lineation is always represented by a point. For large numbers of planes it is recommended to plot the poles rather than the great circles. This makes the image less busy and gives the opportunity to analyze the density of the poles by graphical or mathematical methods. A comprehensive description of the Schmidt net and its applications are given for example in Ramsay & Huber (1983).

For the use of the Schmidt net, place a thin transparent paper atop the net, aligned and tacked at their mutual center. Mark the N direction on the transparency. Suppose that we want to plot the plane 270/66 in the CLAR notation. Turn the transparency until the N mark is lying on the 260 mark of the Schmidt net. Now, using the heavy grid lines, which are spaced 10° apart in the figure in Appendix B, mark the point 66° down (towards the S) counting from the center of the net. You will get the pole of the plane. To get the great circle, rotate the pole on the E-W line. Count 90° over the net center and draw a curved line from N to S through this point.

To plot a lineation, you rotate the transparency counterclockwise on the measured azimuth. Now the plunge angle is drawn from the top (N) of the net towards the center. The resulting point represents the plunge of the lineation.

For the analysis of tectonic data several software programs are available, but the basic principles should be understood to recognize errors or wrong program settings. A typical example of a false interpretation results from wrong basic settings of the program, such as representation on the upper hemisphere, using strike and dip instead of dip direction and dip or wrong allocation of numerical values to linear or planar elements.

Free computer software for the analysis of tectonic data is available from the Department of Geology, Ruhr-Universität Bochum: http://www.ruhr-uni-bochum.de/hardrock/downloads.htm

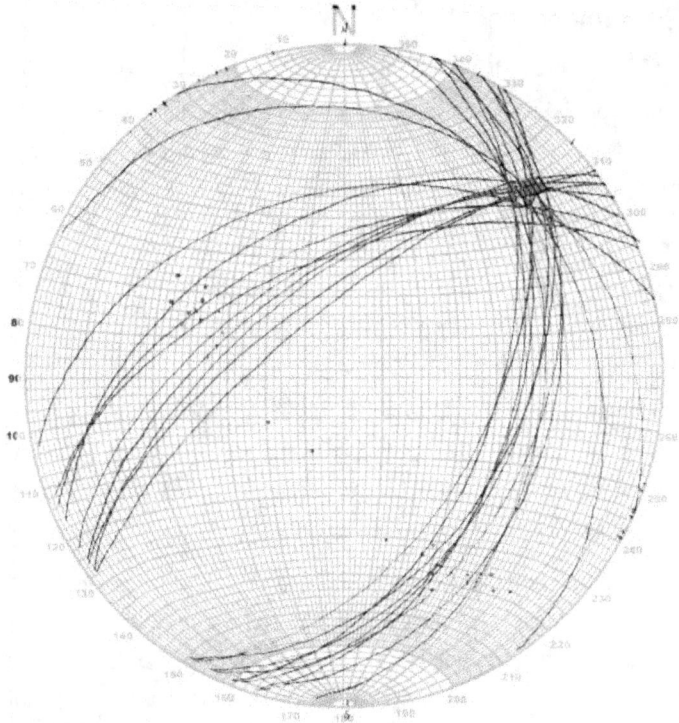

Figure 27: Example for a graphical analysis using the Schmidt Net, lower hemisphere. The great circles and poles of 20 measured planes of a fold structure are plotted. The great circles intersect near the point (048/22), which is equivalent to the orientation of the fold axis.

4. Equipment
4.1 Geological equipment

Students or hobby geologists normally start their career with excursions, led by professional geologists or University teachers. For such one-day trips the necessary minimum equipment comprises:

- hammer,
- hand lens,
- field notebook.

Additionally, safety goggles against rock chips, strong field boots, a daypack with lunch and sufficient drinking water and some kind of rain protection will complete the equipment and will make a geological excursion a successful exercise.

The hammer is the most important and basic tool of the geologist. A robust, hardened steel hammer of ca. 600-800 g weight with a shock-absorbent shaft is required. The shaft length should be at least 25 cm (Fig. 28) to have enough power for cutting a hard rock. It is not necessary to have a quite expensive brand tool (e.g. Estwing®), forged from one piece; a less expensive brick hammer with metal shaft or fiber glass shaft is available in every supplies store (watch out for quality seals!). In no case should a wooden-shafted hammer be used. Wooden shafts break easily or get loose due to shrinking of the wood which may lead to serious injuries. Also cheap fiber glass-shafted hammers bear some risks due to poor manufacturing. Especially when the connection between head and shaft is concealed by a glue sealing, the quality of the connection cannot be controlled.

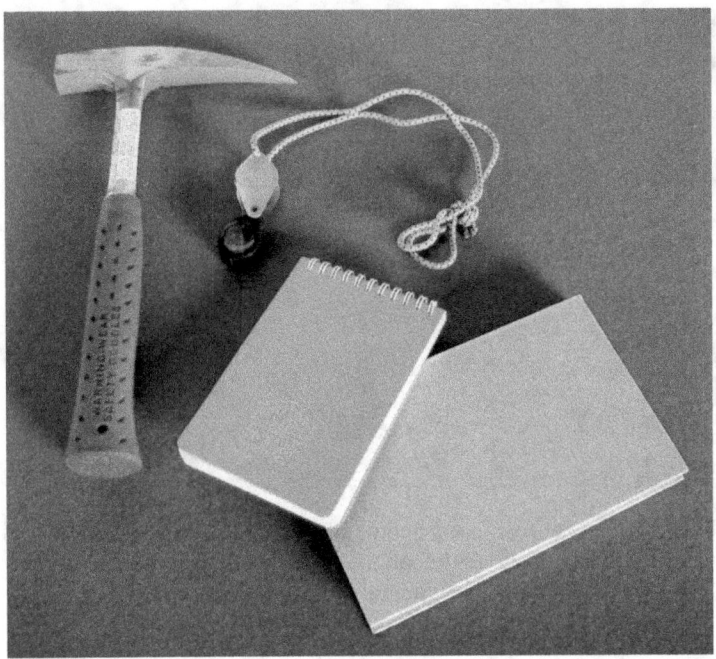

Figure 28: Basic equipment for fieldwork: hammer, hand lens and a water-proof field notebook.

So called jeweler's loupes (Fig. 28) or hand lenses are suitable for geological fieldwork. To identify small minerals or microfossils a 10x magnification is required, and the diameter of the lens should be at least 20 mm. Very important is the quality of the lens; an aplanatic triplet lens which does not distort the marginal field of view would be a good choice. Depending on the casing material (plastic or metal) such lenses are available from 10 US$ upwards from opticians or specialized geology suppliers.

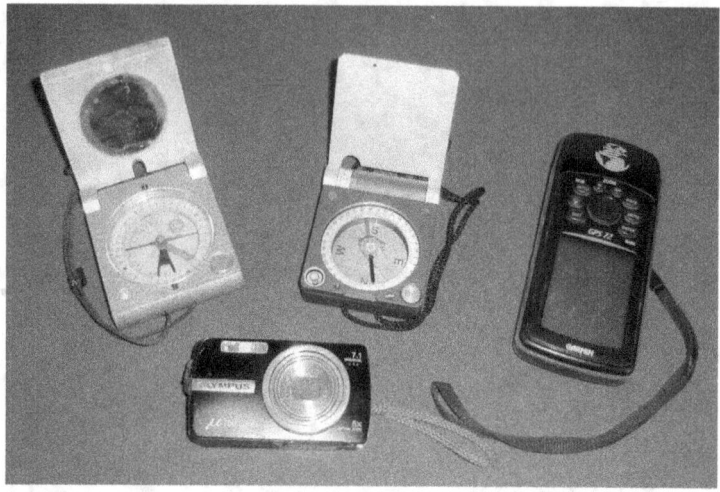

Figure 29: Upper row left, a geological compass type Freiberg, in the middle a Breithaupt® model, on the right side a Garmin® GPS type 72, below a splash-proof digital camera.

A number of special waterproof notebooks are available at special geological supply stores (e.g. *Rite in the Rain*® from the USA). These books are made from waterproof materials (synthetic paper) with fixed or spiral binding. They are quite expensive, however they are extremely useful in a rainy climate where a rain-soaked normal notebook might become useless. For writing, a pencil is recommended since it works everywhere. If you expect a very rainy climate, you may find a special waiter's pencil useful, which is specially designed to write on wet surfaces.

More advanced students and professional geologists need a more comprehensive equipment. This includes:
- a geological compass (Fig. 29),
- topographical and geological maps,

- sample bags,
- marker pens to write on rock samples,
- digital camera.

If required also:
- GPS (Fig. 29),
- sledge hammer for sampling of hard igneous or metamorphic rocks (Fig. 30a),
- barometric altimeter in mountainous regions (Fig. 30b),
- chisel (Fig. 30b),
- magnet and porcelain plate for mineral identification (Fig. 30b),
- hydrochloric acid for testing calcite.

Figure 30: a) Sledge hammer with vertical peen; b) Altimeter, porcelain plate to test the streak, magnet and a small rock chisel.

When a student starts with the first mapping course or a hobby geologists gets interested in tectonics, the questions comes up, which kind of geological compass is suitable. Students should invest in their own instrument to get used to the measuring technique, even outside of University courses. Companies with a large number of geologists should use a single model to avoid confusion. Several fundamental types of geological compasses exist, but the ideal instrument follows the design of Prof. CLAR. This instrument can record dip angle and dip azimuth in one single operation, the two angles which define the orientation of a surface or other tectonic elements in space. The price range between the cheapest instruments of Asian production and a high-end precision instrument is huge, from 75 to approximately 750 US$. For an amateur geologist who wants to record tectonic elements just from time to time, such a low-end instrument could be sufficient. However, a professional geologist who has to record hundreds and thousands of tectonic readings will need an instrument with higher precision and quality. Machining and handling (e.g. magnetic induction damping or a glass bottom for overhead measuring) have their price.

Sledge hammers are available in every hardware shop, but you will find in most cases hammers with two flat ends or a peen perpendicular to the shaft's direction. These tools are not very useful for precise handling, a geological sledge hammer should have a vertical peen. If you cannot find such a tool, you can buy a splitting maul and get the axe-shaped wedge removed by a smith. Such a tool is shown in Fig. 30a.

A digital camera is nowadays a standard device. You can check the quality of the images directly and transfer them on your computer for reporting purposes. If you are planning to purchase a digital camera for geological field usage, the foremost criteria should be the lens quality, protection against splash water, solid manufacturing and battery lifetime, optical zoom and resolution – in this order! Be aware that high resolution gets lost anyway during reproduction or presentation (for example with a projector). Good image quality and a robust casing are more important than 20 Mpixel resolution. Useful is a built-in GPS which allows an automatic recording of the location; in this case only the direction in which the image has been taken, needs to be noted.

For orientation in the field a large-scale topographic map is required. Standard map scales provided by the state surveys in Europe are 1:25,000 or 1:50,000 scale. In mountainous areas local mountain clubs (e.g. Alpine Clubs in some European countries) provide actual map material. As a basis for geological mapping a topographic base map should be enlarged to a 1:10,000 scale. The state surveys often make enlarged black-and-white copies of their map series available on request, especially if the request comes from a University. These enlarged copies should then be cut into a handy format (< A4) and backed with thin cardboard. Geological field observations are recorded on these field maps with colored crayons, traverses are marked by pinholes and outcrops or important notes can be recorded on the back next to the referring pinhole. Always keep a clean copy in your accommodation and transfer your observations daily (!) on to this clean map.

A GPS (global positioning system) navigation instrument is a must for a mapping geologist or a mineral explorer. Even in central Europe, you will frequently find yourself in a situation where an outcrop is located in an area far away from landmarks and the site can be located only with 50 or more meters precision. Nowadays a basic GPS can track a point down to less than 5 meters precision. Unfortunately, the computed altitude is less precise for technical reasons. The altitude on topographic maps is referring to a certain sea level (which is an equipotential plane/geoid). The GPS altitude refers to a mathematical rotational ellipsoid which already has an error to the geoid of ± 50 m. Therefore, a barometrical altimeter, which has to be adjusted at known altitudes several times a day, is still a useful instrument in rugged terrain.

4.2 Safety equipment

Safety First is the first rule for any kind of fieldwork, especially if somebody is working on his own. In such case, always inform a reliable person where you are going and when you are planning to be back. Leave important details like mobile phone number and a copy of your map with the planned route with this person and agree on a time when help should be called. Keep to this schedule or report changes!

Vertical cliffs or overhanging rock walls, especially in old quarries should be avoided. If you enter quarries or work near steep natural cliffs wear a hard hat. Don't climb without a partner and use appropriate equipment and safeguarding. Be aware of dangerous roads; high visibility vests are necessary in all areas with road traffic. In Europe, minimum equipment comprising the following items is recommended:
- Safety goggles (not required for working in soft rocks)
- High visibility vest
- Hard hat
- First aid set
- Mobile phone

If you intend to use a chisel, put a plastic protector on the top end of the chisel and wear protective gloves. Some other protective devices are required in operating quarries and mines (ear protection, special shoes, clothing etc.). In developing countries without mobile phone coverage, a satellite phone is useful. A basic medicine box should be assembled according to recommendation by an experienced physician, who knows the specific situation in your working area. Travel recommendations for other countries are given on the homepages of your Foreign Affairs ministry. Additional security advices can also be found there. One month before you plan an expedition of more than four weeks into a remote area, a medical check is recommended and you should also check with your health insurance if they cover your activities. Consider a special travel insurance, including an emergency evacuation service.

5. Geological Mapping

A geological mapping campaign requires careful preparation. You need equipment appropriate to the geological tasks, the campaign duration and the local climatic requirements. Scientific papers, geological and topographic maps and last but not least permissions from the authorities to collect samples or enter protected areas are necessary. Once in the working area, you should introduce yourself to the local authorities and collect information about access to the working area, which will allow you to draft a working plan. In Europe it might be possible to do this from your desk at home but in developing countries this is normally not possible.

Based on available geological information, the first days should be spent identifying rock units which have been distinguished on previous maps and in publications. You should try to find the freshest outcrops of the different rock units; quarries are the best places for this. After getting an idea of the main lithology, you should traverse a few transects across strike to record main structures of the area. Finally, the area has to be covered, using all available footpaths and by following streams. Mapping progress can be much less than 1 km^2/day in densely vegetated or rugged terrain.

Every observation has to be recorded immediately in the note book, even assumptions. To avoid the loss of your work, a photocopy of your notebook should be made daily or the records have to be transferred into your computer. The content of your field map has to be transferred onto a clean map as well.

5.1 Keeping a field log

Your field notebook is a first order document; interpretation and evaluation of your work are based on these records. You have to keep it in such a way that somebody else can read (and understand) it. Signatures and abbreviations should be shown in a legend at the beginning or end of the book. Don't rely on your memory, always record all observations, even if they don't seem to be very important at the time. It does not matter if you are a student, an amateur geologist, or a hobby fossil-collector, you may find something interesting and the notebook is a document which will allow a scientific evaluation of any discovery.

All records should start day by day with the date, the weather conditions and the way to the first outcrop (don't forget the road conditions). Even experienced geologists neglect to record weather conditions, but the weather has a significant influence on the quality of observations. On a wet and windy day, you are automatically working quickly in an exposed outcrop, making just short notes or the photographs are not good since you tried to protect your camera against rain rather than looking for the best position to make a photograph. If you find later unusual tectonic measurements related to such days, you should go either ignore them or go back and check such measurements.

An outcrop description should be systematic and always follow the same scheme:
- Outcrop number, time, locality (with coordinates); avoid descriptions like *150 meters south of outcrop XY* or *near Z-village* , they are not unambiguous and cannot be tracked back easily;
- Type of exposure (e.g. road cut, quarry, river banks, etc.);

- Rock description (minerals, fabric, name of the rock);
- Stratigraphic allocation;
- Fossils, description, age (if possible);
- Tectonics, geometrical and relative age relations of different rocks in an outcrop;
- Sketches, cross-sections, photographs. Field sketches and cross-sections should always have a vertical and horizontal scale and must be oriented. Photographs should always have an appropriate scale, depending on the size of the object: for example a coin (hand specimen), hammer (small exposure) or a person (quarry or large exposure). Note the direction of view.
- Make notes about collected samples: each sample/fossil requires a unique number.

Digital cameras allow the superposition of date, time and better models with built-in GPS also the coordinates. This is useful for recording and tracking photographs. Make sure that these dates do not overlay the object details.

5.2 Topographic maps for geological field work

A topographic map is a detailed and accurate graphic representation of cultural and natural features of a part of the Earth's surface. The information content and detail depends on the map scale. In Europe official topographic maps with the largest scale are generally cadastral maps of 1:5,000 scale. The next smaller scale is 1:25,000 (small scale = large scale number), which is used as a basic mapping scale. Other common scales are 1:50,000, 1:100,000 und 1:200,000. Limits of official topographic maps are generally bounded by meridians and parallels. Longitude and latitude are commonly printed on the map margin, additionally many topographic maps carry a rectangular grid allowing quick orientation.

Official geological maps are published in different countries at different scales (generally between 1:25,000 and 1:100,000). They are superimposed on simplified topographic maps showing contour lines, roads, settlements and the drainage, commonly in the same pale color.

For geological mapping purposes you can use a black and white copy of an actual topographic map (field slip). The field slip should be at least one magnitude larger than the final scale of the geological map.

Topographic maps are available as a paper copy or in some countries as a digital copy from the land survey (e.g. UK: Ordnance Survey, Germany: State land surveys, USA: USGS, France: Institut Géographique National). In some countries large scale topographic maps are issued by a Military Geographical Service (e.g. Greece) and access for civilians could be restricted. In the countries of the former Soviet Union for example, maps larger than 1:100,000 are classified. The USGS offers a 1:200,000 map series for the territory of the former Soviet Union. If you need more detailed topographic information, you may use aerial photographs or satellite images for your field work.

5.2.1 UTM coordinate system

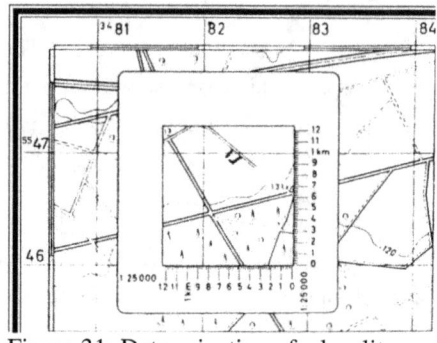

Figure 31: Determination of a locality using Easting and Northing with a ruler.

A rectangular grid is an important addition to a topographic map (Fig. 31), allowing an easy navigation and orientation. Several grid systems are used on topographic maps, the most widespread is the UTM grid. UTM is an abbreviation for Universal Transverse Mercator coordinate system. The system was based on an ellipsoidal model of the Earth. Currently, the WGS84 ellipsoid is used as the underlying model of the Earth in the UTM coordinate system.

The transverse Mercator projection is conformal, so that it preserves angles and approximate shape but invariably distorts distance and area. UTM involves non-linear scaling in both Easting and Northing to ensure the projected map of the ellipsoid is conformal.

The UTM system divides the surface of the Earth between 80° S latitude and 84° N latitude into 60 zones, each 6° of longitude in width and centered over a meridian of longitude. Zones are numbered from 1 to 60. Zone 1 is bounded by longitude 180° to 174° W and is centered on the 177th West meridian. Zone numbering increases in an easterly direction.

Each of the 60 longitude zones in the UTM system is based on a transverse Mercator projection, which is capable of mapping a region of large north-south extent with a low amount of distortion. By using narrow zones of 6° (up to 800 km at the equator) in width, and reducing the scale factor along the central meridian by only 0.0004 (to 0.9996, a reduction of 1:2500) the amount of distortion is held below 1 part in 1,000 inside each zone. Distortion of scale increases to 1.0010 at the outer zone boundaries along the equator.

In each zone, the scale factor of the central meridian reduces the diameter of the transverse cylinder to produce a secant projection with two lines of true scale, located approximately 180 km on either side of, and approximately parallel to, the central meridian (ArcCos 0.9996 = 1.62° at the Equator). The scale factor is less than 1 inside these lines and greater than 1 outside of these lines, but the overall distortion of scale inside the entire zone is minimized.

A position on the Earth is referenced in the UTM system by the UTM zone, and the easting and northing coordinate pair. The easting is the projected distance of the position from the central meridian, while the northing is the projected distance of the point from the equator. The point of origin of each UTM zone is the intersection of the equator and the zone's central meridian. In order to avoid dealing with negative numbers, the central meridian of each zone is given a "false easting" value of 500,000 meters. Thus, any location west of the central meridian will have an easting less than 500,000 meters. For example, UTM eastings range from 167,000 meters to 833,000 meters at the equator (these ranges narrow towards the poles). In the northern hemisphere, positions are measured northward from the equator, which has an initial "northing" value of 0 meters and a maximum "northing" value of approximately 9,328,000 meters at the

84th parallel which is the maximum northern extent of the UTM zones. In the southern hemisphere, northings decrease as you go southward from the equator, which is given a "false northing" of 10,000,000 meters so that no point within the zone has a negative northing value.

5.2.3 Contour lines

A map is a two dimensional object and the third dimension (altitude) is represented by contour lines. A contour line joins points of equal elevation (height) above a given level, such as mean sea level. The contour interval of a contour map is the difference in elevation between successive contour lines.

To maximize readability of contour maps, there are several design choices available to the map creator, principally line weight, line color, line type and method of numerical marking.

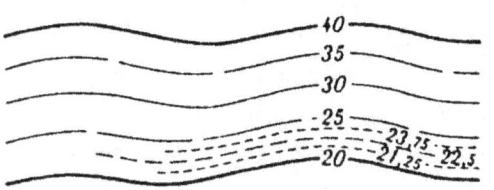

Figure 32: Different line weights, line types and numerical markings occur on a 1:25,000 scale topographic map.

Line weight is simply the darkness or thickness of the line used. This choice is made based upon the least intrusive form of contours that enable the reader to decipher the background information in the map itself. If there is little or no content on the base map, the contour lines may be drawn with relatively heavy thickness. Also, for many forms of contours such as topographic maps, it is common to vary the line weight and/or color, so that a different line characteristic occurs for certain numerical values. For example, in Fig. 32, the 20 m intervals are shown in a different weight from the 10 m intervals.

Line type refers to whether the basic contour line is solid, dashed, dotted or broken in some other pattern to create the desired effect. Dotted or dashed lines are often used when the underlying base map conveys very important (or difficult to read) information. Broken line types are used when the location of the contour line is inferred.

Numerical marking (Fig. 32) is the manner of denoting the altitudes of contour lines. This can be done by placing numbers along some of the contour lines, typically using interpolation for intervening lines. The direction of these text labels is often used to indicate the direction of the slope. Alternatively a map key can be produced associating the contours with their values.

5.2.3 Declination und inclination

Before you go to the field, you have to adjust your geological compass for declination and inclination. The magnetic declination at any point on the Earth is the angle between the local magnetic field —the direction in which the north end of a compass points— and true north (Fig. 33). The declination is positive when the magnetic north is east of true north. Magnetic declination varies both from place to place, and with the passage of time. As a traveller cruises the east coast of Madagascar, for example, the declination varies from 23 degrees west at the southern tip of the island to 11 degrees west at its northernmost point, meaning a compass adjusted at the beginning of the journey would have a true north error of over 12 degrees if not adjusted for the changing declination.

On official topographic maps, a diagram shows the relationship between magnetic north in the area (in most cases for the center of the map at a given date) and true north, with a label near the angle between the magnetic **N** arrow and the vertical line, stating the degree of declination and of that angle, in degrees. Generally, the annual variation is also stated. The magnetic declination will change slowly over time, possibly as much as 2 degrees every hundred years or so, depending upon how far from the magnetic poles it is. This may be insignificant to most geologists, but can be important if using magnetic bearings from old maps.

The angle between magnetic and true north, and the direction from true north to magnetic north is given as a positive value if the declination is lying in a clock-wise direction from true north (Fig. 33). Negative or western declination would indicate that magnetic north lies counter-clockwise from true north.

The values of the declination for any place on the world between 1900 and 2100 can be obtained online from http://www.ngdc.noaa.gov/geomag-web/#declination.

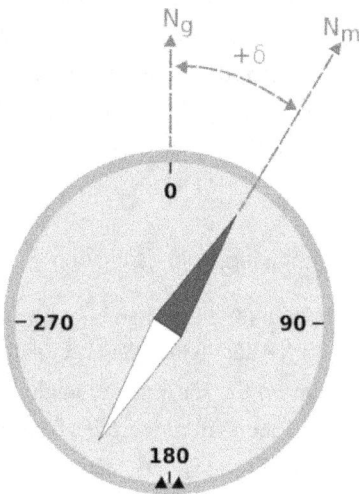

Figure 33: Example of magnetic declination showing a compass needle with a positive (or easterly) variation from geographic north. (image from Wikipedia)

To correct the declination you have to turn the compass scale. Subtract western or negative declinations and add positive or eastern declinations with reference to the index mark of the compass casing.

Magnetic dip or inclination results from the tendency of a magnet to align itself with lines of force. As the Earth's magnetic lines of force are not parallel to the surface, the north end of a compass needle will point downward on the northern hemisphere (positive dip) or upward on the southern hemisphere (negative dip). Contour lines along which the dip measured at the Earth's surface is equal are referred to as lines. The locus of the points having zero dip is called magnetic equator. The range of dip is from -90 degrees to +90 degrees. Without a compensation, the needle in the compass would touch the cover glass and not swing freely. Therefore the southern end of the needle (for a northern hemisphere compass) carries a small weight. If you want to use such a compass on the southern hemisphere, you have to open it and transfer the weight to the other end of the needle. Some liquid-damped, cheaper models cannot be opened, in which case you will need individual compasses for both hemispheres.

5.3 Geological mapping criteria

5.3.1 Outcrops

An outcrop is a natural or man-made place on the Earth's surface where the bedrock is exposed. Typical examples for natural outcrops are
- crags, mountain cliffs, pavements;
- coastal cliffs;
- exposed bedrocks due to landslides.

Common man-made outcrops:
- quarries, pits;

- adits;
- road cuts, tunnels;
- trenches.

In a broader sense a drill hole is also an artificial outcrop.

If a rock is still in situ, we call it a bedrock and it will expose the geological conditions at this location. Very small outcrops do not always allow one to evaluate if the rock is in situ or not. Tectonic measurements from such places have to be taken with great care. If they do not fit into an overall geological picture, ignore such values. In any case, the size and weathering conditions of an outcrop must be recorded in your note book.

5.3.2 Fieldstones

In areas without outcrops (which is most of the Earth's surface) other geological information has to be used. In relatively flat areas without recent periglacial overprint, fieldstones are a useful tool to obtain information about the geology beneath the soil cover. Even they have no connection to the underlying bedrock they represent the partly weathered, local rock type(s), exhumed by tillage or burrowing animals. In central Europe fieldstones account for more than 98% of the area as the primary information source for the underlying bedrocks.

However, in sloped areas they tend to move downhill, the faster with increasing cobble size, slope inclination and substrate humidity. Therefore, the smallest, most angular stones or rock chips are the most reliable source of information. Rock chips and stones of different lithological composition may indicate a lithological boundary, which can be found uphill.

5.3.3 Geomorphology

The morphological features of the Earth's surface reflect the bedrock composition, especially in evolved landscapes with intensive weathering and denudation. Less resistant lithologies form soft landforms, rolling hills and smooth slopes, whereas weathering-resistant lithologies show steep slopes, cliffs and rugged morphologies. Morphological features marking a lithological boundary should be followed and recorded by GPS.

Springs are a very important feature. They are always caused by geological features like faults or underlying, non-permeable lithologies which may help with the interpretation of field observations.

Type and color of soil give important hints for the composition of the underlying bedrock. In tropical areas with thick weathering profiles the soil color may be the only feature helping to identify a lithology between scattered outcrops. Soil color can be recorded with signatures on the map and later transferred into lithological units after some control exposures are found.

In areas without tillage or other forms of agricultural modifications, the natural vegetation can indicate certain elements in the substrate, which then can help to distinguish lithological units.

5.4 Geological maps

A geological map displays rock units or geologic strata shown by color or symbols to indicate surface coverage. Structural features are shown with strike and dip symbols which consist of a long line, a number, and a short line which are used to indicate tilted beds. The long line is the strike line, which shows the true horizontal direction along the bed, the number is the dip or number of degrees of tilt above horizontal, and the short line is the dip line, which shows the direction of tilt.

The basis for a geological map is a simplified, bi-colored (brown-white, grey-white) topographic map. The distribution of different lithologies is printed on top of this map. Tectonic features, such as faults, fold hinges, thrusts and others are shown in solid lines where observed, in dashed lines where assumed. General geological maps show the bedrocks without superficial deposits but special thematic maps exist which focus on other geological topics like composition and thickness of superficial deposits, geochemistry, groundwater resources or tectonic features. For special purposes, a map may not show the Earth's surface but a series of cross-sections.

A geological map has a legend, showing lithologies in chronological order, a cross-section and a stratigraphic profile (the latter is omitted on maps dealing only with igneous or metamorphic units. Each rock unit has a color and an alphanumeric code, which is helpful in colorful maps or for color blind people. All used signatures are explained in the legend. Detailed information about the geology can be found in an explanatory volume accompanying the map.

5.5 Measuring geological structures

Geological information is obtained by recording outcrops and fieldstones on a field map. This gives a two-dimensional map, showing the distribution of different rock types on the Earth's surface. To get an idea about the three dimensional distribution, structural information from all in-situ rocks are recorded and transferred to the field map. We distinguish planar and linear structural elements. Planar elements are bedding in sedimentary rocks, compositional layering in metamorphic and igneous rocks, slaty cleavage and schistosity in metamorphic rocks, faults, thrusts and joints. Linear elements are fold axes, different types of lineations and linear feature in sedimentary rocks like flute casts.

Figure 34: Measuring of a plane by docking the lid on top of a planar rock surface. Rough surfaces can be leveled with an underlying note book.

Such measurement will be carried out by a geological compass. Modern compasses allow the measurement of dip azimuth and dip angle or plunge direction and plunge angle in one single step. Fig. 34 shows the handling of a compass during the measurement. In the compass a small bull's eye level allows the leveling of the casing when the lid is placed on the measured rock surface. Keep the lid on the rock surface and turn the compass casing around the hinge and the center of the lid until the body is leveled. Press the button to release the needle and wait until the compass needle comes to a halt. Release the button and read the azimuth on the graduated circle. The dip angle can be read from the vertical circle attached to the hinge.

Figure 35: Correct measuring a layer from underneath. The azimuth has to be read from the S-pointing end of the needle.

Figure 36: Measuring a lineation by docking the edge of the lid.

With the lid in a position as shown on Fig. 34, the azimuth angle has to be taken from the position of the north-pointing end of the needle. Measuring the bottom side of a layer as shown in Fig. 35 or opening the compass lid more than 180°, the azimuth is taken from the southern-pointing end of the needle.

The measurement of linear structures is similar. The right or left edge of the lid is docked against the linear element and the casing leveled by turning the compass around this lid edge and the hinge into the correct position (Fig 36). Depending on the opening angle of the lid, you have to read the direction at the N-pointing needle (90-180°) or the S-pointing needle (0-90° or >180°).

5.6 Symbols in geological maps

Geological maps are rather complex representations, combining basic topographic information with lithological and structural data. Rock types and rock age are generally shown by different colours, whereas structural information such as bedding, schistosity, fold axes and faults are shown by symbols. Up to now there are no generally accepted standards for these symbols, however some guidelines introduced by geological surveys exist. A comprehensive, trilingual (English, French, German) symbol guide was published by Breddin (1960). This publication also contains a large number of symbols for rock types, fossils and geomorphological features.

A summary of widely used structural symbols is given in Fig. 37. It is highly recommended to copy this figure or a similar one and glue it into the field notebook as a legend. All symbols used in a field notebook and a field map should be explained once.

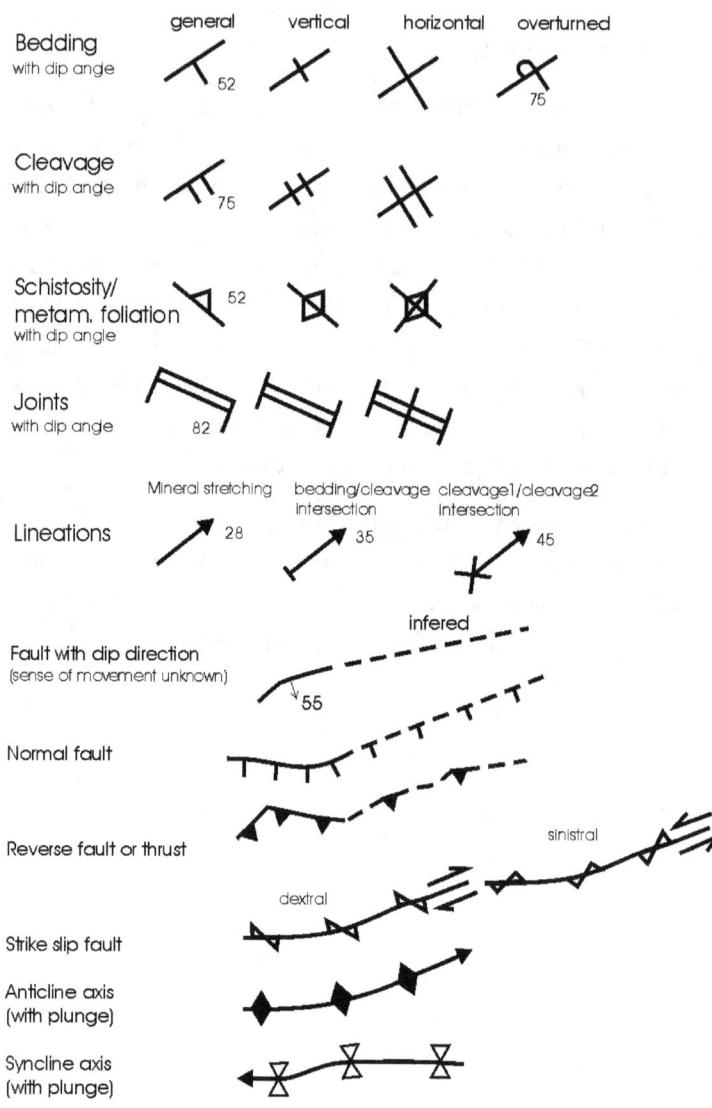

Figure 37: Geological map symbols.

5.7 Collecting samples

One of the most important parts of field work is the selection and cutting of appropriate rock samples, allowing further laboratory investigations. The quality of analytical results mainly depends on the sample quality.

The selection of a sample and its size depends on the scientific problem which you would like to solve. If it is only required to give a precise petrographic description of a rock type, a small piece large enough for a thin section may be sufficient. For geochemical or micropaleontological investigations large volumina may be required. Generally all samples should be fresh, showing no signs of weathering. Quarries are the most suitable places to get fresh material, whereas in natural outcrops large boulders have to be cut in order to get an appropriate piece from the unweathered core. Note the number of the collected sample and the proposed analytical studies in your note book.

The usage of sample identification numbers follows usually different systems, typical are:

1. Successive numbers: 1, 2, 3, ...
2. Outcrop ID/Successive numbers: A5/1, A5/2, A10/1, A10/2 etc.
3. Initials and successive numbers: WB1, WB2 ...
4. Initials+Date+ successive numbers: WB100708/1

For further work, especially in commercial and/or large University laboratories, a short, unambiguous identification number is required. Bear in mind that the identification number may have to be placed on a tiny container or several geologists may drop off their samples at the same time in the laboratory. In particular, the first system of simple numbers is problematic since they may not be unique. The method using an outcrop identification and a consecutive sample number per outcrop has the advantage that a sample can be directly referred to a certain outcrop. The disadvantage is the problem that it is difficult to keep an overview, if all samples are complete, especially if you do not collect samples at each outcrop. If you are working with several colleagues together during a project, an unambiguous sample identification system has to be agreed before starting the field work. In such cases the identification of the sample collector is important since he has recorded the notes referring to the sample, therefore initials plus numbers could be a good tool to identify the samples. Example 4 shows a system introduced in some geology departments. It contains initials of the sample collector, a six digit date and the successive number of samples taken on that date. The only disadvantage might be the relatively long code required for each sample.

Samples should be labeled with water-proof marker pens (working in mafic rocks may require a white enamel pen). If a sample is selected for chemical analyses the labelling should be made on an adhesive tape, but not directly on the rock to avoid contamination. The packing back (plastic, calico) should also be labeled and a sample slip with basic information (sample no., date, outcrop no. and collector) should be placed into the bag. This triple ID system is important because some of the labeling may be abraded during transport. For microfabric analyses or for certain sedimentological analyses an oriented sample might be required. To obtain an oriented sample, a surface (not necessarily the bedding or schistosity, even a plane joint can be sufficient) are measured in situ. The position of the compass lid is indicated by marking a corner of the lid and an arrow showing the direction of the dip azimuth. After cutting the sample carefully out of its position, add the sample number, the measured orientation and the facing of that orientation (top or bottom side) (Fig. 38). This allows the reconstruction of the original orientation. All information is also recorded in the field note book and the sample slip.

Figure 38: Oriented sample, the surface was measured 155/80 on the bottom side of a metamorphic foliation.

It should be noted, that sampling and sample exportation can be regulated by national law. Before starting the field work necessary permits have to be obtained. Generally, such information is available from geological surveys, geology departments at the Universities, customs authorities or board of mines. Getting access and permits for sampling in protected areas generally involves local or national environmental agencies.

References

BREDDIN, H. 1960. *Vorschläge zu einer einheitlichen Darstellung auf lithologischen, tektonischen und hydrogeologischen Zeichnungen und Karten.* Geologische Mitteilungen **1**, 1-120; Aachen.

DUNHAM, R.J. 1962. *Classification of carbonate rocks according to depositional texture.* In: W.E. Ham (Ed.) Classification of Carbonate Rocks, 108-121, Mem. Am. Ass. Petrol. Geol. **1**; Tulsa.

EISBACHER, G. H. 1996. *Einführung in die Tektonik.* 2nd Ed., Enke, Stuttgart.

FOLK, R.L. 1959. *Practical petrographic classification of limestones.* Bull. Am. Ass. Petrol. Geol. **43**, 62-84.

HEITZMANN, P. 1985. *Kakirite, Kataklasite, Mylonite - Zur Nomenklatur der Metamorphite mit Verformungsgefügen.* Eclogae geol. Helv. **78**, 273-286.

LE BAS, M.J. & STRECKEISEN, A.L. 1991. *The IUGS systematic of igneous rocks.* J. Geol. Soc. London **148**, 825-833.

RAMSAY, J.G. & HUBER, M.I. 1983. *The Techniques of Modern Structural Geology.* Vol. 1, p.151-165. Academic Press, London.

PETTIJOHN, F.J., POTTER, P.E. & SIEVER, R. 1973. *Sand and Sandstone.* 617 S., Springer, Berlin.

STRECKEISEN, A. 1967. *Classification and nomenclature of igneous rocks.* Neues Jb. Mineral. Abh. **107**, 144-240.

WINKLER, H.G.F. 1967. *Die Genese der metamorphen Gesteine.* 2nd Ed., Enke, Stuttgart.

Appendix A: Schmidt Net

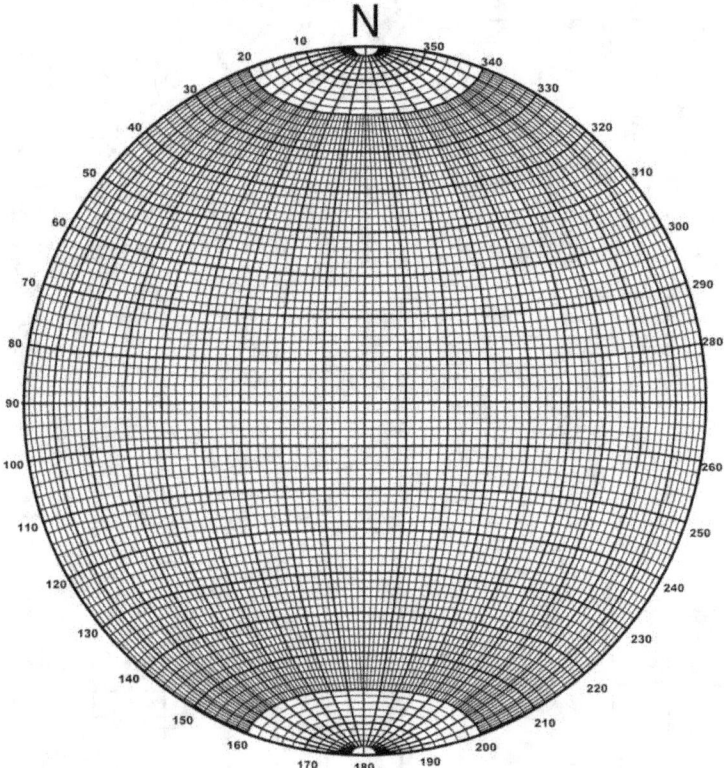

Appendix B: Chronostratigraphic Tables

Time (Ma)	System	Series	Stage	(Ma)	Magnetic Polarity
-0.5-	Quatarnary	Pleistocene	Holocene	0.0117	Brunhes (normal)
			Tarantian	0.126	
-1.0-			Ionian		
-1.5-				0.781	
-2.0-			Calabrian	1.8	Matuyama (reverse)
-2.5-			Gelasian	2.58	

Time (Ma)	System	Series	Stage	(Ma)	Mag. Pol.
	Quatarnary	Pleistocene	Holocene	0.01	
				2.58	
-5-	Tertiary / Neogene	Pliocene	Placenzian	3.6	
			Zanclian	5.3	
			Messinian	7	
-10-			Tortonian		
				11.6	
		Miocene	Serravallian	14	
-15-			Langhian	16	
-20-			Burdigalian	20	
			Aquitanian	23	
-25-		Oligocene	Chattian	28	
-30-			Rupelian		
-35-				34	
	Tertiary / Palaeogene		Priabonian	38	
-40-			Bartonian	41	
-45-		Eocene	Lutetian		
-50-				48	
			Ypresian		
-55-				56	
		Paleocene	Thanetian	59	
-60-			Selandian	61.6	
			Danian	66	

Time (Ma)	System	Series	Stage	(Ma)	Mag Pol.
-70-	Cretaceous	Late C.	Maastrichtian	66	Rapid Polarity Changes
				72	
-80-			Campanian		
			Santonian	83.6	
-90-			Coniacian	86	
			Turonian	90	
-100-			Cennomanian	94	
				100.5	
-110-			Albian		
				113	
-120-		Early C.	Aptian		
				126	
-130-			Barrêmian	131	
			Hauterivian	134	
-140-			Valanginian	139	
			Berriasian	145	
-150-	Jurassic	Late J.	Tithonian	152	
			Kimmeridgian	157	
-160-			Oxfordian	164	
			Callovian	166	
-170-		Middle J.	Bathonian	168	
			Bajocian	170	
			Aalenian	174	
-180-			Toarcian		
				183	
-190-		Early J.	Pliensbachian	191	
-200-			Sinemurian	199	
			Hettangian	201	
-210-	Triassic		Rhätian	209	
-220-		Late T.	Norian		
				228	
-230-			Karnian	235	
-240-		Middle T.	Ladinian	242	
			Anisian	247	
-250-		Early T.	Olenikian	251	
			Induslan	252	

Time (Ma)	System	Series	Stage	(Ma)
		Lopingian	Changsingian	252
260			Wuchiapingian	254
	Permian	Guadalupian	Capitanian	260
270			Wordian	265
			Roadian	269
280			Kungurian	272
		Cisuralian	Artinskian	279
290			Sakmarian	290
			Asselian	296
300		Pennsylvanian	Gzhelian	299
			Kazimovian	304
310	Carboniferous		Moskovian	307
			Baschkirian	315
320				323
330		Mississippian	Serpukhovian	331
340			Viséan	
350			Tournaisian	347
360				359
370	Devonian	Late D.	Famennian	372
380		Middle D.	Frasnian	383
			Givetian	388
390			Eifelian	393
400		Early D.	Emsian	407
410			Pragian	411
			Lochkovian	419

Time (Ma)	System	Series	Stage	(Ma)
		Přidoli		419
420		Ludlow	Ludfordian	423
	Silurian		Gorstian	426
430		Wenlock	Homerian	428
			Sheinwoodian	431
		Llandovery	Telychian	433
440			Aeronian	439
			Rhuddanian	441
			Hirnantian	443
450		Late O.	Katian	445
			Sandbian	453
460			Darwillian	458
470	Ordovician	Middle O.	Dapingian	467
			Floian	470
480		Early O.	Tremadocian	478
490			Stage 10	485
		Furongian	Jingshanian	489
500			Paibian	494
			Guzhangian	497
510			Drumian	501
	Cambrian		Stage 5	505
			Stage 4	509
520			Stage 3	514
530		Tereneuvian	Stage 2	521
540			Fortunian	529
				541

Note: Some series and stages of the Cambrian have not been named yet.

Time (Ma)	Eon	Erathem	System	(Ma)
600	Proterozoic	Neo-	Ediacarian	541
800			Cryogenian	635
1000			Tonian	850
1200		Meso-	Stenian	1000
1400			Ectasian	1200
1600			Calymmian	1400
1800		Paleo-	Statherian	1600
2000			Orosirian	1800
2200			Rhyacian	2050
2400			Siderian	2300
2600	Archaean	Neo-		2500
2800				2800
3000		Meso-		
3200				3200
3400		Paleo-		
3600				3600
3800		Eo-		
4000				4000
4200	Hadean			
4400				

Appendix C: Comparison Charts

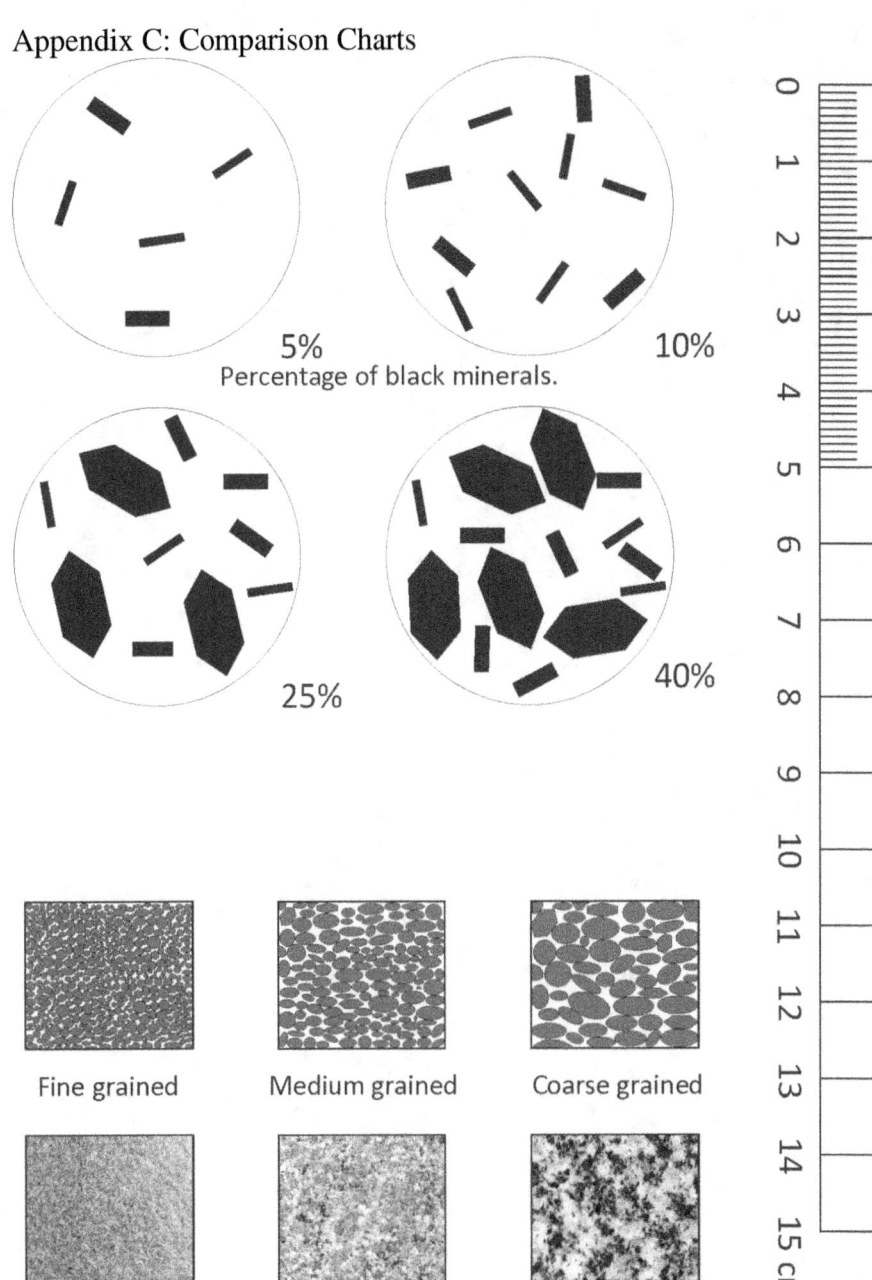

5% 10%
Percentage of black minerals.
25% 40%

Fine grained   Medium grained   Coarse grained

Notes:

www.ingramcontent.com/pod-product-compliance
Lightning Source LLC
Chambersburg PA
CBHW050112230526
45470CB00004B/1791